Molecular Biology in Narrative Form

BERKELEY INSIGHTS IN LINGUISTICS AND SEMIOTICS

Irmengard Rauch
General Editor

Vol. 64

PETER LANG
New York • Washington, D.C./Baltimore • Bern
Frankfurt am Main • Berlin • Brussels • Vienna • Oxford

Priya Venkatesan Hays

Molecular Biology in Narrative Form

A Study of the Experimental Trajectory of Science

PETER LANG
New York • Washington, D.C./Baltimore • Bern
Frankfurt am Main • Berlin • Brussels • Vienna • Oxford

Library of Congress Cataloging-in-Publication Data

Venkatesan Hays, Priya.
Molecular biology in narrative form: a study of the experimental trajectory of science / Priya Venkatesan Hays.
p. cm. — (Berkeley insights in linguistics and semiotics; v. 64)
Includes bibliographical references.
1. Molecular biology. 2. Narration (Rhetoric). I. Title.
[DNLM: 1. Molecular Biology. 2. Writing.
QH 506 V461m]
QH506.V464 572.8—dc22 2006022367
ISBN 0-8204-8699-X
ISSN 0893-6935

Bibliographic information published by **Die Deutsche Bibliothek**.
Die Deutsche Bibliothek lists this publication in the "Deutsche Nationalbibliografie"; detailed bibliographic data is available on the Internet at http://dnb.ddb.de/.

29 Broadway, New York, NY 10006
www.peterlang.com

For Alain Cohen
my hope and spirit

Contents

Figures

Acknowledgments

This book is the result of varied academic conversations with Alain Cohen of the Department of Literature at University of California, San Diego. As my primary dissertation advisor, he guided me in nurturing fashion to articulate the ideas expressed in this work. I can only hope that this study reflects his brilliance and methods. My thanks also go to Christopher Lowrey of the Department of Medicine at Dartmouth Medical School, whose advice and encouragement were invaluable in the writing of this book. I would also like to thank Ana Minvielle, administrative coordinator at UC San Diego's Literature department, who lent her support in a most gracious fashion. Thanks also go to my family and friends, and most especially, my husband, Ezra Hays.

Priya Venkatesan Hays
Hanover, New Hampshire
September 2006

Introduction

> The problem of signification is at the center of the preoccupations of our time...The human world as it appears to us is defined essentially as the world of signification. The world can only be called "human" to the extent that it means something. Thus it is in research dealing with signification that the human sciences can find their common denominator. Indeed, if the natural sciences ask questions in order to understand how man and world are, the human sciences pose question, more or less explicitly, of what both of them signify.
>
> —Algirdas Julien Greimas, *Structural Semantics: An Attempt at a Method*: 1

When narratologists talk of the beginnings of their discipline, they typically refer to the publication in year 1966 of issue number 8 of the French periodical *Communications* (Jahn 2003: 196). Although the term narratology had yet to be coined, the title of the special issue succinctly anticipated what narratology would be all about—*L'analyse structurale du récit*, the structural analysis of narrative. Introducing the subject in the lead essay ("Introduction à l'analyse structurale des récits"), Roland Barthes begins by presenting his famous list of story forms:

> There are countless forms of narrative in the world. First of all, there is a prodigious variety of genres, each of which branches out into a variety of media, as if all substances could be relied upon to accommodate man's stories. Among the vehicles of narrative are articulated language, whether oral or written, pictures, still or moving, gestures, an ordered mixture of all those substances: *narrative is present in myth, legend, fables, tales, short stories, epic history, tragedy, drama[suspense drama], comedy, pantomime, paintings (in Saint Ursula by Carpaccio, for instance), stained-glass windows, movies, local news, conversation.* Moreover, in this infinite variety of forms, it is present at all times, in all places, in all societies; indeed narrative starts with the very history of mankind; there is not, there has never been anywhere, any people without narrative; all classes, all human groups, have their stories, and very often those stories are enjoyed by men of different and even opposite cultural backgrounds: narrative remains largely unconcerned with good or bad literature. Like life itself, it is there, international, transhistorical, transcultural. (Barthes 1975: 237, italics mine)

Nowhere in this assessment of narrative does Barthes mention science or scientific discourse. The presumption of this text, however, is that science, particularly molecular biology, can be presented in narrative form and that molecular biology can serve as an object of narratology. In other words, the empirical approach of laboratory investigation can be analyzed in light of its demonstrated narrative properties. The purpose of this book is to apply a conception of narrative discourse in literary theory in order to illuminate the construction of a particular scientific argument/text in the field of transcriptional biology and the routines, procedures, methods, and conventions that surround laboratory work and research. The category of molecular biology that I employ to make some of the central arguments of this study I have chosen is the use of the experimental method. This category I define precisely throughout the text as the materials and methods used in making discoveries in modern transcriptional biology. The premise of my argument is that these procedures and methods are organized narratively, which I characterize throughout the text. This interpretive method applied to a case study in molecular biology may be used to interpret similar sorts of research.

The aim of this text is to explore one aspect of scientific communication from the perspective of literary theory. My main questions arise from my primary work on literary theory and my interest in scientific writing. I focus on language and interpretation, and argue that literary theory contributes to our understanding of scientific writing.

I have chosen narratology as the model for an analysis of the scientific process because it has become an interdisciplinary subject. Fields as varied as film theory, literary studies, and cognitive science have considered narrative as an applicable analytical tool for the investigations of questions of style, point of view, and interpretation (Chatman 1993: 1; Martin 1986: 4; Herman 2003: 19). We usually assume that rational knowledge is based on a single set of principles which, when applied to particular subjects of study, produce different theories and structural models. Once freed of the notion that [narratologists] should study only stories that are untrue and highly respected (the domain of traditional literature), critics realized that the anthropologist, the folklorist, the historian, and even the psychoanalyst and the theologian are all concerned with narratives in one way or another (Martin 1986: 23). Thus, stories based on reality became the object of narratology. In turn, I would include the scientist as part of these categories because narrative, as I will elaborate upon, marks a pivotal characteristic of science in which its process can be conceived of as having a structure. The structural basis of narrative offers a powerful paradigm for assessing how molecular biology perpetuates itself, since, as David Gooding argues, research processes have been reconstructed into the orderly stream of narratives (Gooding 1992: 65). In addition, as Karin Knorr-Cetina states: There is, in [molecular biology] laboratories, another memory organ that preserves experience. This is the narrative culture of the laboratory—more specifically stories. Stories are scenarios of former experiences that participants have had directly or have heard about. They are told and retold, on appropriate occasions, and thus travel through the laboratory. As long as they circulate, they keep the relevant experience alive, turning it into a communal stock of knowledge (Knorr-Cetina 1999: 106).

As I will argue later in the text, the narrative analysis of molecular biology allows for a construction of novel conceptual frameworks to conceive of science. A narrative is an event, and scientific endeavors are a manifestation of natural events. In other words., scientific statements, as I will explore, demonstrate their sustainability through their representation as events. In this manner, narratology is a privileged field for the examination of the various forms of scientific communication.

Furthermore, since time is a fundamental component of biological processes, including molecular biology, narrative analysis offers a potent examination of the characteristics of science. Uli Margolin constructs a dissection of narrative in terms of temporality. He explores storytelling in terms of retrospective, concurrent and prospective narratives. (Margolin 2001: 1). According to Porter-Abbott, narrative is the principal way in which our species organizes its understanding of time. Narrative is so much a part of the way we comprehend the world in time that is virtually built in to the way we see (Porter-Abbott 2002: 6). The inevitable linearity of story makes narrative a powerful means of gratifying the need for causation, which is an integral element of the biology of transcriptional events in gene expression. In short, narratology offers a method of dissecting how discovery in the regulation of gene expression occurs, and of allowing valid claims that concern the nature of the scientific project to emerge. Event, time and causation are inevitable properties of both narrative and science.

The programme of narrative analysis made giant steps forward through the work of A.J. Greimas and the Paris School of Semiotics, which included the likes of Julia Kristeva, Roland Barthes, Gerard Génette and Claude Lévi-Strauss. The analysis provided by semiotic analysis allows for a research apparatus that can specifically attribute a structural form to almost every human endeavor. In this study, I focus on what narratology has to offer to molecular biology in terms of how its rhetorical narratives are focused on modalities. Françoise Bastide, a scientist herself, precisely brought to bear on how narrative analysis is predicated on modalities.

There are other endeavors that explicitly address the scientific text and the processes of laboratory investigation and enable the use of a linguistic method that lies at the heart of the scientific endeavor. However, narratology, since it is predicated essentially on time and causation, remains a most appropriate model to construct a paradigm of the methodological properties of biological statements and procedures. Change and causality are endemic to the processes of molecular biology, and laboratory method and experiment reflect these key parameters. Much of the molecular biological discovery I highlight and analyze is subsumed under the trajectory of experimental time and causation, properties of narrative as discussed earlier. For example, a routine biological statement such as a gene is said to be transcribed evokes a temporal implication that events of agency are set in motion, and that these events cause the activity of transcription. Narrative concerns the chronology of events and causation since questions such as origin of life on earth are answered narratively.

That is, the methods I employ attribute a methodological depiction of the communicative apparatus at work in the scientific endeavor, particularly in molecular biology. This methodological characteristic is inherent as a consequence of how the methods in transcriptional and chromosomal biology are employed, how their objects are determined and known, and how their terms are formulated and then put forth in a formal publication. The elucidation of the signification of these processes is formidably advanced by the semiotic project.

The concept of scientific production as narrative has been articulated. Steve Woolgar shows that the facts and objects of scientific inquiry are retrievable in a five-step narrative of the relationship of texts and so-called objects of scientific knowledge (Schleifer 1992: 30): (1) "Science" begins when an investigator merely writes down or somehow records potential relations and possible facts in a working document. (2) On this basis, the investigator is in a position to focus and clarify the potential object of inquiry through manipulation and revision within or among documents. (3) At the point at which the object of inquiry becomes well defined, the investigator easily, and perhaps "naturally," moves to posit the independent existence of the object being studied. (4) The investigator can and will clarify the inquiry further by elevating the object of inquiry to an existence both independent of and, eventually, even prior to the initial documents. From this point forward, the object of inquiry can be said to be only "reflected" or noted in the documents. (5) In the final stage of consolidating investigation and discovery, it is incumbent on the investigator—to remain consistent with empirical and scientific protocols—to deny the existence of steps (1), (2), and (3) (Woolgar 1988: 68).

This formulation makes a crucial point concerning the textual manifestation of scientific objects of inquiry and their renunciation in the following of scientific procedure. Identifying the alleged contradictions inherent in the scientific process, Woolgar notes, that as an object of inquiry is subsumed under laboratory investigation, the cognitive apparatus that puts it forward is obviated. However, this assessment is somewhat problematic for a number of reasons that underscores the premises of this book. The category of the events of the scientific process as delineated by the specific narrative elements that Woolgar highlights unnecessarily distinguishes between the initial construction of scientific objects of inquiry and the methodology that gives rise to scientific output, indicative of the production of these objects. In other words, the textual discourse of science is not an empirical part of the procedure. My particular method is predicated upon the presumption that rather than both these activities being in contradistinction to each other, one in fact can be derivative of the other. My argument is based on the notion that the scientific endeavor evokes the elements of narrativity. Woolgar's assessment also characterizes science as being composed of contrasting parts, rather than as being an entity whose features complement each other in a unitary fashion. The essence of narratology is that the entities of storytelling, which I argue includes scientific discourse, form a structure, or related units that compose a whole; as I will argue, science is an entity of storytelling.

For Woolgar, the object of inquiry is the experimental method, which he translates into scientific discourse. This particular translation is addressed in Chapter 5, i.e., the transformation of scientific knowledge into scientific discourse. However, the constructed model I assume, which I will explore later, is that the various objects of inquiry in this analysis is subsumed under biological fact through its articulation in written scientific publications, and as a culmination of the scientific process.

Science as manifested in laboratory experimentation is characterized by narrative programs that are at the heart of its persuasive capacities, highlighting the transition from *faire-savoir* (knowing) to *faire-croire* (believing). This transformation was articulated by Françoise Bastide (*La Démonstration, Documents*, 1981: 5):

> Un article scientifique constitue un tel discours sans récit englobant précisant la situation de communication; toutefois, avant de tenter de décrire des "effets de sens" de faire-savoir et de faire croire en analysant un article à titre d'exemple, il paraît nécessaire de delimiter quelque peu les champs du croire et du savoir dans ce domaine particulier. La description des controversies scientifiques par les sociologies[1] fournit les elements de la situation de communication scientifique: production d'articles par un laboratoire,

[1] Here, Bastide is referring to B. Latour and S. Woolgar, *Laboratory Life: The Social Construction of Scientific Facts*, London, Sage, 1979.

> reception et utilization par les autres laboratories. Chaque article y est présenté comme un "coup" (au sens qu'a ce mot dans un jeu) et la réussite, ou l'échec, est mesurable par le nombre et la durée des citations d'un article dans les publications d'autres laboratoires. L'intérêt de l'information communiqué repose donc sur sa valeur d'usage, en termes sémiotiques. L'avantage d'une telle presentation du texte scientifique est de mettre l'accent sur le caractère persuasive de la publication, et d'éviter toute reference à la vérité, qui, dans le domaine des sciences expérimentales, est chose très labile (Bastide, 1981: 6).

> A scientific article constitutes such discourse without narrative including precisely the situation of communication; however, before attempting to describe the "effects of meaning" of knowing and of making believe in analyzing an article from its title by example, it seems necessary to delimit somewhat the fields of believing and knowledge in this particular domain. The description of scientific controversy by sociologists provides the elements of the situation of scientific communication: the production of articles by a laboratory and their reception and utilization by other laboratories. Each article presents there like a takeover, an imposition (in the sense of a play on words), and the success and failure are measured by the number and length of citations of an article in the publications of other laboratories. The interest of the communicated information is therefore built on its worth of usage, in semiotic terms. The advantage of such a presentation of the scientific text is to put emphasis on the persuasive character of the publication and to avoid all reference to the truth, which in the domain of experimental science is labile (Bastide, 1981: 6, Translation mine).

In this essential set of statements, Bastide associates crucial terms within the narrative endeavor vis-à-vis scientific communication. Meaning is articulated in the frame of a persuasive apparatus that demarcates the transition from knowing to believing among the laboratory community. Persuasion is the hallmark of any scientific publication, the model here, and to decipher the mechanism of how such a persuasion (or convincing) takes place is the project of semiotics. Bastide introduces the concept of semiotics as underscoring the project of scientific communication and its persuasive characteristics. Yet, Bastide makes the qualification that the scientific process is tantamount to being a rhetorical endeavor that is not directly related to the acquisition of wisdom or knowledge, since it has dubious access to truth. Acknowledging that the "success" of any scientific project is contingent upon citations in other publications, Bastide points out that it is linguistic usage that is at stake here. Albeit her framework, the power of the semiotic apparatus will become apparent in the assessment of the narrative properties of the scientific text.

Bastide continues by developing the context of scientific controversy:

> Il faut bien remarquer toutefois que l'utilisation des travaux de sociologie dans une aire culturelle particulière, et qui nous a conduit à distinguer savoir et croire sur le base due critère de l'absence ou de la presence d'une controverse, a des consequences directes

> sur la manière dont les "effets de sens" de faire-savoir et faire-croire seront identifiés dans les articles scientifiques: la dimension syntagmatique sera nécessairement privilégiée par suite d'une attention particuliére portée aux anti-programmes et aux anti-sujets qui paraissent propres à traduire, á l'intérieur du discours, la presence ou l'absence d'une situation polémique (designee comme "controverse" à l'extérieur) (Bastide 1981: 7).

> It is necessary to remark however that the use of the work of sociology in a particular cultural area, and this brings us to distinguish knowledge and belief on the basis of the criteria of the absence or presence of a controversy, has direct consequences on the manner in which the "effects of meaning" of knowing and making believe are identified in a scientific article: the syntagmatic dimension will be necessarily privileged by paying particular attention brought to the anti-programs and anti-subjects that seem appropriate to translate, within discourse, the presence or the absence of a polemic situation (designated as outside "controversy") (Bastide 1981: 7, Translation mine).

Bastide brings up central issues in assessing the narrative program of science; her analysis lies at the heart of the paradigms that are utilized in attributing biology in narrative form in terms of narratology and its derivation in semiotics. The polemical situation of a narrative, of which anti-subjects and anti-programs are a part, constitutes the fundamental concept in the semiotic apparatus of the elementary structure of signification, the semiotic square based on the binarism of categorical terms: A:B :: -A:-B (Greimas 1966: 13). This elementary structure, on which our humanity rests, fulfills the requirement that in order for a narrative to have meaning, a narrative must form a signifying whole and "thus is organized as an elementary semantic structure." (Hawkes, 1991: 92). This analysis necessarily entails that the persuasive properties of a text, as articulated by the transition between knowing and making believe, are inherently defined by natural oppositions. A scientific term evokes its binary in establishing its argument. The essential insight of the text is thereby generated. Thus, the ultimate signification of a text is based on difference.

The implications of the narrative paradigm for molecular biology are that the terms inherently raise distinctions, i.e., the trajectory of polemics lies at the heart of the scientific text. Without the oppositions evoked by the terms, meaning cannot be generated from the text. Anti-programs and anti-subjects, when revealed, form the components of a narrative programme that can provide the basis of how the text generates its significative properties and its persuasive capacities.

The approach to the scientific text taken here is along analogous lines. I consider the scientific process from its perpetuation in the empirical method to its production in scientific publication as objects for narratological analysis. This thesis focuses on the intersection where French narrative theory and molecular biology intersect creating a new hybrid of textual analysis that might be called "molecular narratology." This notion of bridging science and narratology has been evoked before, particularly in the edited anthology *Narrative Theory and Cognitive Science*, in which the term "cognitive narratology" was coined. Cognitive narratology is based on a body of work that interprets narrative phenomena as essentially cognitive processes of perceiving and mapping information through specific mental processes (Herman 2003: 19). However, I have based my methods upon the Greimasian paradigm, to which Bastide refers, that states that the binary structures of language, i.e. the manifest differential character of language, are intimately related to the nature of meaning of any text. As my narratological endeavor extends to scientific enunciations and their correlative practices and theories, my analysis consists of a metatheory, or a theory of a theory. Rather than thinking of the interplay as yet another interdisciplinary topic, the importance of my analysis rests on the realization that the reciprocity of science and the humanities obliterates all traditional disciplinary boundaries (Ormiston and Sassower 1989: 14). Reciprocity takes place between these boundaries: if such a reciprocity is to be appreciated, it must be seen as an erasure of the boundaries and the intellectual segregation of disciplinary inquiry. (Ormiston and Sassower 1989: 14). I wish to merge the interdisciplinary boundaries between molecular biology and literary theory throughout this text.

Extending this project entails a more precise description of the implications of the Greimasian paradigm: i.e. my assessment that scientific discourse rests on operations of negation and assertion that involve semantic constructions based on the differences of terms that can be dissected semiotically as narrative objects. Greimas' articulation of this relies upon his characterization of narrative subjects involved in confrontation and struggle:

1. The existence of two subjects S1 and S2 (or of a subject and an-anti-subject), which correspond to two contradictory doings, with the relationship of contradiction being, as we know, an unoriented relation;
1. The semantic restriction of the syntactic doing through establishing equivalence between the operation of *negation* and the function *of dominance*, the result of polemic antagonism (italics Greimas, Greimas 1987: 74).

Underlining the indefinable nature of the lexeme (a word considered as a unit) reminds us of [occupations] when, in inquiring into the conditions in which signification appears, we were led to postulate:

1. That no object is knowable in and of itself. Only by its determinations can it be known;
2. That its determinations could be apprehended only as differences etched against the object, and this differential meaning gives to these determinations the status of linguistic value (Greimas 1987: 86).

The notion of biological information based on difference has been articulated by Hoffmeyer and Emmeche. According to Gregory Bateson, information is based on difference. A sensory end organ is a comparator, a device which responds to difference...Genes are important for understanding the text of DNA, as words are for literature. But to understand what genes mean, semantics and hierarchy of context must be consulted...With the birth of molecular biology and especially molecular genetics in the 1950s and 1960s,...suddenly a new foreign vocabulary was introduced into biology, that of information theory. Terms like 'program,' 'genetic code,' 'information,' and 'messenger RNA,' 'feedback,' and the like became respectable and even indispensable notions. Facilitated by a widespread indifference to epistemological problems among biologists, the concept of genetic information became for all practical purposes, identified as the sum of genes which carried it...Genes must be considered as signs, not as particles or "pieces of DNA"...What we propose, then, is that the traditional paradigm of biology be substituted by a semiotic paradigm the core of which is that biological information is understood primarly as sign (Hoffmeyer and Emmeche 1991: 136).

Hoffmeyer and Emmeche proceed to conduct a semiotic analysis of the molecular evolution in terms of the Peircian triad and of what they describe as the code-duality nature of language in terms of the digital and analog. However, in the context of Greimas' statements, the assessment made earlier concerning scientific discourse can be further expanded upon. Negation and affirmation in science is the rubric upon which the knowledge of the object comes into being, and this knowledge is derivative of the antagonistic relationship of the linguistic units through which it is articulated. This knowledge is mediated through determination that arises from "differential meaning" as the result of the polemical nature of the discourse. The inherent duality of negation and affirmation that exists in any linguistic framework, of which the scientific text is an example, demarcates the ontology of the objects in the text. The scientific text puts forward a set of statements, for which the Greimasian apparatus reveals their determinability. As I will show in chapters 3 and 4, the semiotic project highlights how certain scientific objects are known through the discourse of binaries generated in the text.

In this manner, I have chosen the semiotic project to discern the processes of molecular biology, with its varied objects of inquiry, including a written article and its perpetuation as scientific fact. The semiotic project in terms of molecular biology asks: What is the relation between DNA and the "programme"? How does this scheme take into account the external logic of the programme itself? And above all, who is the subject to whom this programme makes sense? (Hoffmeyer and Emmeche 1991: 118). I interpret these questions to indicate that research on DNA is a laboratory practice immersed in a narrative logic, which the semiotic project highlights. These questions are relevant for this narratological application of molecular biology since the programme can be considered as a narrative schema.

Semiotics is based on the inherent binary nature of language, i.e. word units naturally form oppositions and these oppositions lead to differential meaning of a story or narrative. This paradigm of narrative analysis, based on semiotics, can be applied to scientific investigation and involves a precise, sequential analysis of scientific enunciations as pronounced in a scientific text. In this context, Julia Kristeva remarks:

> *Ce projet fait de la sémiotique un mode de pensée, une méthode qui pénètre aujourd'hui toutes les sciences socials, tous les discours scientifiques ou théoriques concernant les modes de signifier et occupe la ligne même où se joue la lutte entre science et ideologie. Remplaçant la philosophie classique, la sémiotique devrait être la théorie scientifique de l'époque dominée par la science.* (Kristeva 1971: 4).

> This project makes of semiotics a mode of thought, a method that penetrates today all the social sciences, all scientific discourse or theories concerning the modes of the signifier and occupy the same line or plays itself the struggle between science and ideology. Replacing classical philosophy, semiotics should be a scientific theory in an epoch dominated by science (Kristeva 1971: 4, Translation mine).

All application of a scientific theory demands efforts and particular research, and in the present case of the application of the theory of formal languages and the analysis of natural language, a specific development of formal languages in the sense that is agreed upon (*Toute application d'une scientifique théorie exige des efforts et des recherches particulières, et dans le cas présent de l'application de la théorie des langues formalisées à l'analyse de la langue naturelle, un développement spécifique des langues formalisées dans le sens qui convient*, Paduceva 1971: 177, Translation mine).

Nonetheless, we can ask whether recognition of the bi-isotopic nature of discourse is generalizable. We can go even further and say that from this perspective, science can be interpreted as an effort to pierce and see through the seeming of common sense in order to reach its being-true. This view presents science as a victory of immanence over manifestation (Greimas 1987: 177). This amounts to recognizing and accepting the need for a fundamental distinction, an *apparent* level of narration, at which the diverse manifestations of narrative are subject to the specific requirements of the linguistic substances through which it is expressed, and an *immanent* level, which is a kind of common structural trunk where narrativity is located and organized at the stage preceding its manifestation. Therefore the *linguistic structures* of narrative correspond, at the level of manifestation, to *narrative structures*, and the analysis of narrative has discourse analysis as its corollary (italics Greimas, Greimas 1987: 64).

This preceding analysis implies that the statement that "science is a victory of immanence over manifestation" predicates the existence of narrativity independent of its linguistic characteristics. In others words, the narrative of science, precisely speaking, is about the methodological structure of its discourse rather than intuitive articulation. This entails that an analysis that reveals the structural properties of a text, such as semiotics, can illuminate the method behind the rhetoric of a text.

Chapters 3 and 4 deal directly with methods of narrative analysis as applied to scientific pronouncements and how the properties of the scientific method inure itself to refinement through a theoretical assessment of its plural nature. (cf. Barthes, Chapter 2, 3). As I argue on the polemical nature and plurality of the scientific text, the inevitable conflict and controversy that arises in science are assessed by a narrative programme.

Semiotic analysis allows for a sociolinguistic construction of knowledge, and as this analysis is permitted in a scientific field (*champs de la scientifique*), semiotics can elucidate the cultural content of the expression plane (Eco) of science:

> *En effet, la sociologie de la connaissance comme l'analyse du langage, sans perdre de vue le fait evident que le sujet connaissant est toujours un individu concret,, considèrent celui-ci comme un être socialement formé et conditionné, et ces deux approaches se rencontrent sur le plan commun des conditionnements sociaux de la connaissance humaine. On ne peut contester le caractère social de ces deux analyses, dont*

> *l'une concerne les influences exercées par les interest d'un groupe social sur la vision que l'individu a de la réalité, tandis que l'autre se rapporte aux influences du langage socialement formé sur la pensée individuelle, à commencer par la perception et l'articulation du monde qu'impose un système linguistique donné...Il me semble utile d'étudier tant les points de junction de la sociologie de la connaissance et de l'analyse du langage sur le plan de l'objectivité de la connaissance...L'analyse de ces rapports permettra de mieux comprendre le problème même de l'objectivité de la connaissance, mais aussi d'approfondir l'analyse sociologique du problème grâce à l'implication de son aspect linguistique, et réciproquement l'analyse linguistique grâce à l'implication de son aspect social.* (Schaff 1971: 63-64)

> In effect, the sociology of knowledge like the analysis by language, without losing sight of the fact that the knowing subject is still a concrete individual, considering which like a socially formed and conditioned being, and these approaches meet together at a common point of social conditions of human knowledge. One can contest the social character of these two analyses, of which one concern is the influence exercised by a social group on the vision the individual has of reality, while the other relates to the influences of language socially formed on the individual thought, and begins by the perception and articulation of the world that imposes a given linguistic system...[It seems] useful to study such points of junction between the sociology of knowledge and the analysis of language at the point of objectivity of knowledge...The analysis of these rapports will permit us to better understand the problem likewise of the objectivity of knowledge but allow to go deeper into the sociological analysis of the problem thanks to the implication of its linguistic aspect (or semiotic aspect), and reciprocally thanks to the implication of its social aspect (Schaff 1971: 63-64, Translation mine).

Schaff highlights the relationship between linguistics and sociology that Greimas developed in his theoretical models on sociolinguistics (see Chapter 2). Social signification must be sought in natural languages. In other words, natural languages become the signifiers that enable us to identify and contrast social groups with regard to their sense of whether they belong to linguistic communities. In this case, sociolinguistic significations are the signified whose signifier is made up of natural languages and their articulation (Greimas 1987: 193).

In turn, Thomas Sebeok makes the broad assertion that science has been permeated with semiotic issues all along; clearly all the sciences as well as the arts are bridged, whether by single conduit or by network configuration, through the semiotic paradigm. This is the vital contribution of semiotics, namely, that it entails the inevitable marriage of the arts with the sciences because, as soon as either is touched by man, through his willful choice of a given method, the same human factor influences or interprets the [science and art] in identical ways (Sebeok 2000: 77).

The scientific phenomena being analyzed are within the rubric of biological frameworks that encompass the range of the scientific endeavor. In each chapter I treat science from the perspective of its attributes as experimental method

(Chapter 3), a written article (Chapter 4), and scientific discourse (Chapter 5). The premise of this project is that the assumption that representation of biological event possesses narratological properties; and, that biological objects must be assessed with this narrative method as the model of inquiry.

However, the justification for applying narrative analysis to molecular biology should rest on the understanding that I am treating science from these varied trajectories. Through the engagement of precise analytical moves, I will explore how these diverse conceptual fields that demarcate biology are illuminated and conceived of in a novel fashion in terms of how the scientific process and text are narratively structured.

In Chapter 3, I cite Porter-Abbott's narrative analysis of the theory of evolution in the context of my own segmentation (see Chapter 3) of scientific statements and articulate its relevance. In Chapter 4, I cite Lily Kay's semiotic analysis of the genetic code and, as well, ascribe the pertinence of her analysis. These narrative and semiotic treatments of aspects of the natural sciences point to the tradition of applying such perspectives to science and the consensus among many life scientists and writers that molecular biology can be the focus of an interdisciplinary study in this manner. The originality of my own analysis stems from two basic factors: (1) a blow-by-blow dissection of a review article in transcriptional regulation, vis-à-vis Greimas; (2) the focus on French narrative theory in terms of statements and procedures in molecular biology.

Chapter 1

An Explanation of the Processes of Molecular Biology Used As Objects of Analysis in this Text

The trajectory of this book is structured with the intent of reaching audiences in literature and in the humanities. This objective is contingent upon an explanation of the scientific processes that are discussed in the following chapters in the context of narrative and semiotic theory. These processes must be clarified and elucidated for the humanities reader in order that he or she understands the significance of merging French theory with molecular science.

The type of science treated here is molecular biology and genetics, with emphasis on DNA and chromosome structure and dynamics. In this chapter, each genetic statement and molecular biological process that I have articulated will be elaborated upon and described in order to make it accessible to the diverse audiences I am trying to reach. I will begin with a discussion of the structure of the gene and regulation of gene expression and then of chromosome structure and dynamics and describe how they are integrated and delve into the experimental techniques that were employed in crucial discoveries in these areas. I will end with an assessment of why I have considered these biological processes in the context of narrative and semiotic theory and sociolinguistics.

The objectives of narratology are to help understand, analyze, and evaluate narrative. A narratologist presumes that the text is a finite, structured whole composed of signs. A narrative text is a text in which an agent relates a story in a particular medium, such as language, imagery, sound, buildings, or a combination thereof. A story is a *fabula* that is presented in a certain manner. A fabula is a series of logically and chronologically related events that are caused or experienced by actors (Bal 1997: 5). The term "story" includes the events, characters and setting, as well as their arrangement. Discourse is the means by which the content is communicated, or what the form of expression is (Martin 1986: 108).

A narratologist uses the necessary tools to find the structural relations among the various elements of a narrative. In turn, the geneticist focuses on how living systems function at the cellular, or molecular, level and studies the growth, structure, development and general characteristics of microorganisms and cell lines. As I will argue, integrating these two objectives entails a presupposition that the diverse elements of scientific phenomena, text and discourse, belong to the corpus of narrative texts. As I will demonstrate in subsequent chapters, narratological systems that have been developed can be applied to the process of discovery in molecular biology and generate insight into the mechanism of the scientific endeavor and the nature of its productive activity.

One example of this merging of narrative theory and scientific processes is the concept of catalyses. Catalyses is a concept that Barthes developed that indicates events in a story that are not necessary for the perpetuation of the story. Catalyses can be compared with explicit and implicit cardinal functions that are necessary for the progression of a narrative. One can see an analogy with RNA production in that, as I will show, there exist various processes that may not be integral or pivotal to the discovery related to transcription from its starting point to end.

The Structure of the Gene and the Central Dogma

Genes are DNA, deoxyribonucleic acid. DNA is a double helix consisting of antiparallel strands in which the nucleotide units are linked by 5'→ 3' phosphodiester bonds. The nucleotides are adenine, thymine, guanine and cytosine, which are arranged on the double helical backbone. These nucleotides are called pyrimidines and purines, or pyramidal and purinic bases (Figure 11). The **central dogma** defines the paradigm of molecular biology. Genes are perpetuated as sequences of nucleic acid, but function by being expressed in the form of proteins. The **genetic code** describes the relationship between the sequence of DNA and the sequence of protein. The genetic code is a series of three nucleotides that **translates** the bases of the DNA sequence into the amino acid sequence of proteins, which are made of amino acids.

A chromosome consists of an uninterrupted length of duplex DNA that contains many genes. Each gene is **transcribed** into an RNA product, which in turn is translated into a polypeptide sequence when the gene codes for a protein (Lewin 2000: 35).

Gene expression occurs by a two-stage process: Transcription generates a single-stranded RNA (ribonucleic acid, containing adenine, uracil, guanine and cytosine bases) identical in sequence to one of the strands of the duplex DNA. The main RNA generated is called messenger RNA, or mRNA. Translation converts the nucleotide sequence of RNA into the sequence of amino acids comprising a protein. An mRNA is translated into a protein sequence. The entire length of an mRNA is not translated, but each mRNA contains at least one coding region that is related to a protein sequence by the genetic code; each nucleotide triplet (codon) of the coding region represents one amino acid.

One can immediately see the relevance of narratology here in terms of its objectives as the transcription of the genetic code is an analogy for the process of narrative.

Transcription and the Regulation of Gene Expression

Transcription, the process whereby DNA is converted into RNA, occurs in a multi-step process. It is at the level of transcription in the process of the conversion of DNA into protein that most of the regulation of gene expression takes place. By having acknowledged that control of gene expression can occur at multiple stages, and that production of RNA cannot inevitably be equated with production of protein, it is clear that the overwhelming majority of regulatory events occur at the initiation of transcription (Lewin 2000: 649-50). The cardinal function of gene expression is a function of processes that occur at the beginning; the middle stages are catalyzers that do not affect the ultimate initiation of transcription.

The term "regulation of gene expression" denotes the concept whereby the production of the protein products of genes are controlled by factors that influence the activity of the gene. The concept can be compared to an assembly line of buses that are stationary but standing bumper to bumper with some space between the bumpers, representing genes composed of DNA bases on a string. As people board a particular bus, this initiates the process of setting the bus in motion. However, the bus can only move if a driver boards and starts the engine. As one driver accelerates, a bus moves, analogous to the initiation of the process of transcription. While this analogy is not entirely precise (a gene is transcribed into a product in the process of activation rather than instigating a process of mobility), it indicates how regulation of gene expression is under the control of extragenetic factors that temporally and spatially affect its activity.

Transcription occurs in three stages: initiation, elongation, and termination. Transcription is initiated on a DNA sequence called a promoter, which precedes a gene; the promoter is termed as being upstream of a gene, or occurring before the gene. The best-known promoter is called the TATA sequence. This sequence is essential because the enzyme (a protein that accelerates chemical reactions) that transcribes a DNA sequence (the gene), RNA polymerase, elongates the RNA sequence after the TATA binding protein binds to the TATA promoter to initiate transcription. RNA polymerase is a large enzyme composed of multiple subunits. As the TATA binding protein binds to the TATA sequence, the two DNA strands melt and unwind to allow the RNA polymerase to attach RNA nucleotide units to the DNA bases: Adenine is paired uracil, guanine with cytosine. For example, if the DNA sequence on the gene is TCCGTA, the messenger RNA molecule is UAACAT, which is composed of two codons. As RNA polymerase moves down the DNA molecule, basepairing RNA nucleotides to it, a transcription bubble forms, which is basically two separated DNA strands that allow for RNA polymerase to enter and transcribe.

A unifying principle is that transcription factors have primary responsibility for recognizing characteristic sequence elements of any particular promoter, and they serve in turn to bind the RNA polymerase and to position correctly at the startpoint (Lewin 2000: 646). RNA polymerase cannot initiate transcription itself, but is absolutely dependent on auxiliary transcription factors (Lewin 2000: 627). (Figure 12). Together with these factors, the enzyme constitutes the basal transcription apparatus that is needed to transcribe any promoter (Lewin 2000: 627). Here, we see the purpose of narratology apparent with the goals of molecular biologists. The process whereby these auxiliary factors initiate transcription temporally and spatially, and distinguished further in terms of how they cause transcription, is the object of much laboratory research. Exploring this in terms of narrative analysis is an endeavor that succinctly characterizes the elements of laboratory investigation, as I will discuss.

The first step of transcription apparatus formation at a promoter containing a TATA box is binding of the factor TFIID to a region that extends upstream from the TATA sequence. TFIID consists of types of component; the TATA-binding protein and other subunits called TATA-binding protein associated factors. TFIID is then joined by TFIIA and TFIIB to allow for the RNA polymerase to bind DNA and initiate transcription. The elements of time and causation, the property of any narrative are inherent in this chronological and spatial account of transcription initiation.

Regulation of Transcription

The principle that emerges from characterizing groups of genes under common control is that they share a promoter element that is recognized by a regulatory transcription factor. An element that causes a gene to respond to such a factor is called a response element: examples are heat shock response element, glucocorticoid response element and serum response element. Response elements have the same general characteristics as upstream elements such as promoters and may sometimes be located upstream. We assume that all response elements function by the same general principle. A gene is regulated by a sequence at the promoter or enhancer that is recognized by a specific protein. The protein functions as a transcription factor needed that activates RNA polymerase. Active protein is available only under conditions when a gene is to be expressed (Lewin 2000: 650).

The glucocorticoid receptor is an example of a prominent regulatory transcription factor which, after the receptor's activation by the steroid hormone glucocorticoid, binds to a specific response element where it activates transcription.

Transcription factors contain a variety of structural motifs that interact with specific DNA sequences. Transcription factors typically are classified according to the type of DNA binding domain they contain. One example are the

Homeodomain proteins. (see Figure 13 for structure of DNA binding domain). Transcription factors with this type of DNA-binding domain are called homeodomain proteins, a name derived from a group of genes in *Drosophila* in which, when mutated, lead to transformation of one body part to another during development.

A second example are the zinc-finger proteins: These proteins surround a zinc ion that bind to DNA-binding domains. They are termed "zinc fingers" because a two-dimensional diagram of the structure resembles a finger. Many are proteins that bind with high affinity (high potency) with steroid hormones, such as the glucocorticoid receptor mentioned above.

A third example are the leucine-zipper proteins: These transcription factors dimerize with DNA binding domains through leucine residues. Leucine is an amino acid (a component of proteins) that is hydrophobic in nature (not attracted to water).

Nucleosomes and Chromosome Structure

Chromatin has a compact organization in which most DNA sequences are structurally inaccessible and functionally inactive (Lewin 2000: 657). This means that DNA is harbored in a sequestered form in an organization in which the DNA sequences are not directly accessible for the process of transcription. DNA is packaged into chromosomes in a "beads on a string" format. The fundamental subunit of chromatin is the nucleosome (Figure 14). The nucleosome contains approximately 200 base pairs of DNA, organized by an octamer of small, basic (above pH 7 due to the protein's amino acid content) proteins into a bead-like structure. The protein components are histones, which form the interior core. The DNA lies on the surface of the histone particle.

The second level of organization is the coiling of a series of nucleosomes into a helical array, and this is followed by a third level of organizational packing in which this helical array is packaged into fibers. We need to work through these levels of organization to characterize the events in transcription. We assume that association with additional proteins, or modification of existing chromosomal proteins that are involved in changing the structure of chromatin. Transcription requires unwinding of DNA, and thus must involve an unfolding of the structure that allows the relevant enzymes to manipulate the DNA. This is likely to involve changes in all levels of organization (Lewin 2000: 567).

The nucleosome is approximately 200 base pairs of DNA associated with a histone octamer, or set of 8 histone proteins. This consists of two copies of H2A, H2B, H3 and H4, the core histones. The DNA, associated with the core histones, is called core DNA. There is an additional histone protein termed H1 in which is wrapped linker DNA, which sits in a somewhat perpendicular position and attaches to the core histones. DNA associated with the histones is called packaged DNA. DNA not associated with any of the histones is called naked DNA.

When chromatin is digested with the enzyme micrococcal nuclease, the DNA is cleaved into integral multiples of a unit length. One can determine the size of this DNA using a technique known as gel electrophoresis. The results of the procedure reveal that the DNA is fragmented into discrete sizes that correspond to approximately 200 base pairs in length. This is significant because in subsequent chapters I describe experiments that assess regions of DNA accessible to transcription factors through the use of micrococcal nuclease. A proportion of the DNA is specifically attacked and thereby indicates susceptible regions.

In Chapter 3 I describe a scientific controversy that involves the biological phenomena of **nucleosomal positioning**. What is nucleosomal positioning and how is it relevant to the laboratory experiments I outline? This question is answered by asking more questions: Does a particular DNA sequence always lie in a certain position with regard to the topography of the nucleosome? Or are nucleosomes randomly arranged on DNA, so that a particular sequence may occur at any location, for example in the core region in one copy of the genome and in the linker region in another? (Lewin 2000: 586).

To investigate these questions, one must use a defined sequence of DNA, or determine the position relative to the nucleosome of a defined point in the DNA (see Figure 15). The term nucleosome positioning means that a nucleosome is associated with a unique sequence of DNA in a particular configuration, so that each site on the DNA is always located at particular position on the nucleosome.

The account of nucleosomal positioning offers an analogy to narrative, because as I will discuss later, it is intimately linked to the debate surrounding the movement of chromatin during transcriptional regulation.

Figure 15 outlines an experiment called indirect end-labeling that demonstrates nucleosomal positioning. The enzyme microccocal nuclease discussed earlier digests the nucleosome into fragments of DNA. Next, the fragments are digested with what is known as a restriction enzyme. A restriction enzyme is a biological enzyme that cuts the DNA at specific DNA sequences. Let's say you have a restriction enzyme for the DNA sequence ACTGTC. A restriction enzyme specific for this sequence will cut the DNA sequence at a particular point called the restriction site there in the sequence. In indirect end-labeling, the restriction enzyme cleaves at the target sequence. When determining the size of the resulting DNA fragment after this set of digests, a single band indicates that the nucleosome has a unique sequence of DNA.

Figure 16 outlines a scenario whereby nucleosomal positioning is absent. In this case, the linker DNA (also known as linkers) consist of different DNA sequences in each copy of the genome; so the restriction site lies at a different DNA sequences in each copy of the genome. Therefore upon restriction enzyme digestion, multiple bands, shown as a smear, will result.

Chromosome Structure and Regulation of Gene Expression

"Chromatin Structure and Cardiac Gene Expression," a paper I address in Chapter 4, discusses how chromatin is accessible for transcription in the context of **hypersensitive sites**. What is a hypersensitive site and how is it relevant in this context? Chromatin consists of DNA wound around a series of nucleosomes, more loosely packaged in regions that contain active genes (Lewin 2000: 593). In addition to the general changes that occur in active or potentially active regions, structural changes occur at specific sites that are associated with initiation of transcription or with certain structural features of DNA. These changes were first detected by the effects of digestion with the enzyme DNaseI.

When chromatin is digested with DNaseI, the first effect is the introduction of breaks in the DNA duplex at specific, **hypersensitive** sites. These sites represent sites where DNA is exposed, that is, not organized in the nucleosomal structure. Again, the technique of indirect end labeling described earlier with consecutive digests by micrococcal nuclease and DNAaseI generates hypersensitive sites.

Many of the hypersensitive sites are related to gene expression (Lewin 2000: 593). Every gene has a site, or sometimes more than one site, in the region of the promoter. Most hypersensitive sites are found only in the chromatin of cells in which the associated gene is being expressed. The 5' hypersensitive site (5' means that the DNA strand begins with a 5' phosphate molecule) appears before transcription begins, and the DNA sequences contained within the hypersensitive sites are required for gene expression. This was revealed by mutational analysis in which the bases within these hypersensitive sites were mutated, and the protein product was not visible from gels generated by the technique of Western blots. This is consistent with the conclusion since the protein product would be evidence of gene expression as dictated by the central dogma of DNA→ RNA→ protein.

In the context I have elaborated above, "Chromatin Structure and Cardiac Gene Expression" focuses on how gene expression (when transcription factors are made accessible to the gene) takes place when chromatin is actively remodeled, that is when its nucleosomes move. This "nucleosomal perturbation," as the paper's authors describe it, allows for regions on DNA to be made accessible to transcription factors. The regions on DNA are termed *cis-acting* sequences while the transcription factors are termed *trans-acting* factors. The paper deals specifically with cardiac gene expression, and the authors found a transcription factor called the GATA factor whose promoter was surrounded by two cardiac-specific DNAase hypersensitivity sites in the upstream region of this promoter. GATA was shown to be integral in overcoming the chromosomal barrier to transcription (as described earlier) since these DNase hypersensitivity sites allow for the GATA factor to bind to its site over nucleosomal DNA. The conclusion of the paper is that chromatin is actively remodeled in this process.

The cellular DNA is packaged into, as described earlier, into nucleosomes. Expression of a gene requires changes in the state of chromatin: the essential issue is how transcription factors gain access to the DNA of the promoter (Lewin 2000: 666). The debate over nucleosomal positioning that I describe in

Chapter 3, and that is at the heart of Chapter 4's analysis of "Chromatin Structure and Cardiac Gene Expression," is that histones can be viewed as "repressing" or inhibiting transcription; thus chromatin is a passive structure. However, it may be the case, as articulated by the paper, that interactions between transcription factors and chromatin can be required for activation.

As outlined in Chapter 3, the debate about nucleosomal positioning in initiating the process of transcription hinges on whether nucleosomes are precisely positioned at some response elements or that nucleosomes move to allow for transcription factors to bind DNA. This argument is made in the context of what is known as histone acetylation. (An acetyl molecule is a small carbon-based molecule.) The pertinent point about histone acetylation is that by attaching the acetyl molecule to a histone, activation of gene expression occurs and can indicate if transcription takes place whether the nucleosomes move or not.

In this assessment, the goals of narratologists can be fruitfully merged with the objectives of molecular biologists. As narratologists hope to show the underlying laws of literary structure, biologists hope to discover the underlying descriptions of natural phenomena. Both utilize models in their determinations. Biologists use models to determine if nucleosomes move or not. Narratologists utilize syntactic and semantic models to categorize the coherence of stories. The trajectories are analogous in terms of the emphasis on structural causation and temporality.

The focus of the next chapter are the narratological models of French narrative theory that serve to characterize the narrative nature of how biologists make these underlying discoveries and how the scientific process is perpetuated through narrative models. My first premise is that the scientific process is in essence narrative in nature. Narrative is a complex term; one way I am conceiving of narrative is considering it as an object that one can characterize as a pattern of binaries, and constructed through a dissection that ultimately ascribes a definite structure to that object. There is a rhetorical and linguistic element to the production of scientific investigation, which will be one of the objects of my analysis. Some of the questions I will try to address are:

(1) How is methodological self-awareness among molecular biologists enhanced?

2. How can we reconceive and rethink scientific enunciations through the precise analytical moves of narrative theory? Here, scientific "fact" is the object of analysis. In this context, I explore the rela-

tionship of the method to discovery, and elucidate the narratological properties of scientific statements.

3. How is the rhetorical apparatus of scientific publications revealed through an analysis of the text, particularly through Greimas' semiotics of text (See subsequent chapter for explanation). The scientific text, in this case, a review article, was chosen to highlight the communicative apparatus that engenders its persuasive capacity.
4. How can we reassess scientific controversy and conflict, as expressed through data, by epistemologically characterizing molecular biology through textual frameworks that sought to characterize contradiction?

These central aspects that I address are related since the components treat how the ultimate results of scientific investigation, observational data and interpretation and their articulation in publication, can be objects of analysis for uncovering the linguistic parameters of scientific statements and their potential methodological implications.

Chapter 2

An Explanation of the Principles of Narrative Theory Used as Analytical Tools in this Text

In the previous chapter I attempted to articulate a representation of molecular biology that is accessible to non-scientific audiences. However, I treat science in the context of systems of thought in the humanities, which, in turn, may not be accessible to scientists, who may be unfamiliar with terms in narratology and semiotics. In this chapter, I will introduce and describe the concepts of narrative analysis and semiotics as developed by Roland Barthes and Algirdas Julien Greimas.

This analysis is relevant for the field of molecular biology because genetic discovery involves a structural method that lies at the heart of narratology. The *fabula*, understood as material or content that is worked into a story, has been defined as a series of events. This series is constructed according to certain rules. This series is termed the logic of events. Structuralists [in French narrative theory] often work from the assumption that the series of events that is presented in a story answers the same rules as those controlling human behavior, since a narrative text would otherwise be impossible to understand...An event...always takes up time. Events, actors, time and location make up the elements of a story, which produce a desired effect, whether aesthetic, convincing, etc. (Bal 1997: 7). The goal of narratology is the ordering of various elements of story. My objective is to transpose this goal onto molecular biology.

In this context, I consider science within the confine of two separate trajectories:

(1) How scientific statements concerning transcriptional regulation and chromosomal activity are assessed through the model of narrative discourse;

(2) How scientific statements achieve their rhetorical effect by demonstrating how scientific conclusions are reached through the framework established by textual analysis.

The objects of inquiry in this study are the procedures and methods of molecular biology. This chapter will be organized around the French theorists Roland Barthes and A.J. Greimas whose intellectual range includes literature, philosophy, history and cultural studies. As a literary theorist, I employ the concepts articulated by them with special reference to the implications of their work for literary texts. This is essential because writing and representation are at the heart of the scientific enterprise; Barthes' statements about *ecriture* may be significant for communication in molecular science. After this discussion, I will specifically elaborate on the importance of Barthes' and Greimas' analyses in uncovering the linguistic mechanisms and implications of molecular biology.

Roland Barthes, the Codes of Narrative and the "Writerly" Text

Consider Barthes' opening statements on the literary text from *S/Z*, an analysis he wrote of the Balzac's short story "Sarrasine": "Because the goal of literary work (of literature as work) is to make the reader no longer a consumer, but a producer of the text. Our literature is characterized by the pitiless divorce which the literary institution maintains between the producer of the text and its user, between its owner and customer, between its author and reader. This reader is plunged into a kind of idleness—he is intransitive ... reading is nothing more than a referendum" (Barthes 1977: 4). Opposing this "readerly text," which is what is read but not written, is the "writerly text" in which the reader is actually constituting in the act of active writing. Writing is essential to the process of reading and interpretation involves recognizing the plurality of a text. Plurality designates a text which is the locus of more than one meaning and hence generative of different but legitimate readings (Sturrock 2003: 118).

Denotation is the logic or grammar of the text, based on "truth, objectivity and law" (Barthes 1977: 7). Connotation indicates the plurality of the text. Therefore the terms "denotation" and "connotation" refer to the readerly and writerly text respectively. To Barthes, everything in a text is a sign, or is connotative. Barthes' narrative analysis and his method is the *explication du texte,* at the heart of which is connotation: "Then, what is connotation? Definitionally, it is a determination, a relation, an anaphora, a feature which has the power to relate to other parts of the text...Analytically, connotation is determined by two spaces: a sequential space, a series of orders, a space subject to successivity of sentences, in which meaning proliferates by layering; and an agglomerative space, certain areas of the text correlating other meanings outside the meaning of the text...(Barthes, *Image Music Text* 8). The concept of connotation is relevant for the Barthesian narrative apparatus since "Semiologically, each connotation is the starting point of a code, the articulation of voice which is woven into the text" (Barthes 1974: 9). The codes, described below, which Barthes formulates and uses to dissect "Sarrasine" are the basis for his narrative analysis.

The relevance of this analysis for the molecular biologist is that within every statement derived from procedural observation and experiment there exists a connotative plurality. The objective of narrative theory generates insight into the molecular biological endeavor since Barthes' evaluation allows for science to be analyzed intertextually.

In other words, simple denotative statements generate an infinite set of meanings; there is something to be said about each statement or set of statements. Literature owes its existence to the codes that we invent to process the world and to create it (Hawkes 1991: 112).

There are five main codes:

(1) The hermeneutic code: This consists of "all units whose function is to articulate in various ways a question, its response, and the variety of chance events which can either formulate the question or delay its answer; or even constitute an enigma and leads to its solution" (Barthes 1974: 17). This really is the "storytelling" code, by means of which the narrative raises questions, creates suspense and mystery (Hawkes 1991: 116). The hermeneutic code groups together the various sequences of the text that first pose and then resolve the "enigmas," so creating suspense in the reader (Sturrock 2003: 119).

(2) The code of semes or signifiers; the semic code: This is a code of connotations that uses hints or "flickers of meaning" generated by certain signifiers (Barthes, 19, cited in Hawkes 1991: 117). The semic code groups together various recurrent connotations or themes of the text (Sturrock 2003: 119).

(3) The symbolic code: This is the code of recognizable "groupings" or configurations, regularly repeated in various modes and by various means in the text. By pointing out the antithetical nature of certain passages in the text, the symbolic codes build up to a climax where the notion of antithesis (such as male/female) enters the overall meaning of the text.

(4) The proairetic code, or the action code: This is the code of actions. Derived from the concept of proairesis, "the ability rationally to determine the result of an action," this code in embodied in such sequences as "I was deep in one of those daydreams" (from "Sarrasine") implies a state of absorption that will be brought to an end, indicating something

will happen to change that state. The action code groups sequences of actions forming self-evident wholes, however scattered they may be textually (Sturrock 2003: 119).

(5) The referential or cultural code: This code groups all the many references made in the text to the cultural, social, and other knowledge of the author's day (Sturrock 2003: 119).

It is Barthes' purpose to show that the structures of a text like "Sarrasine" are literary, that is an elaborate work of literary art, not fictionalized social observation, and the codes serve as method for this elaboration and refer to the connotative aspects of a narrative (Sturrock 2003: 120). However, the system of codes refers to only the aspect of narrative analysis. In attempting to assimilate alternative theories of narrative (Tzetevan Todorov and A.-J. Greimas), he articulated a framework by which narrative generates meaning through functions, actions and narration. He systematically set up an apparatus that dissects the components of narrative down to its discrete levels of language, function (units) and actions, and in addition articulated a system of narrative, as will be described. As he himself remarked, "Thus, in order to describe and classify an infinite number of narratives, a 'theory' (in this pragmatic sense) is needed and the immediate task is that of finding it, of starting to define it." So begins the "structural analysis of narrative" (Barthes 1977: 82). In order to conduct a structural analysis, according to Barthes, it is thus first of all necessary to distinguish several levels or instances of description and to place these instances within a hierarchical (integrationary) perspective (Barthes 1977: 86). What Barthes means by description is that each narrative is composed of units, i.e. phonemes which are integrated into words, which are then integrated into sentences. Only in this integration can meaning be generated. Description refers to systems of meaning generated by levels of organization rather than to a heterogenous mass of elements. (This is reminiscent of the hierarchical organization of chromosome structure). Barthes offers a paradigm from which a descriptive-based science such as molecular biology can be transformed into a model in which narrative properties are determined.

For Barthes, the fundamental significance of functions was how it developed a determination of units. In a word, we have to define the smallest narrative units (Barthes 1977: 88). By functional, Barthes means that everything in narrative, down to the smallest detail, has meaning. Functions refer to one of the two main classes of units, indices being the other class. Functions involve metonyms; while, indices invoke metaphors. To understand the difference, take the statement "The telephone rings." Explicitly the telephone is ringing (functional); implicitly however, someone picks up the telephone (indices). Function corresponds to the functionality of doing, while indices correspond to the functionality of being. Both are necessary for meaning to be generated.

There are two subclasses of functional units, the cardinal unit and the catalyzer. For a function to be cardinal, it is enough that the action to which the function refers opens an alternative that is of direct consequence to the subsequent development of the story; in short, that the function inaugurate or conclude an uncertainty. It is the structural framework of cardinal functions which accomplishes the telescoping of logic and temporality (Barthes 1977: 94). Cardinal functions ascribe to narrative the quality of clarity in following the logical thread of a narrative; it could be considered the denotative aspect of the story. It is an essential property in attributing to narrative the characteristics of being able to construct a causal chain of events in a logical fashion within the thrust of the narrative. Catalyzers, although present in narrative, are not necessary for its perpetuation. Both cardinal functions and catalyzers are necessary in scientific discourse, as I will describe.

Language proper [*langue*] can be defined by the concurrence of two fundamental processes: articulation or segmentation (form), which produces units, and integration, which gathers these units into higher levels of units (this being meaning). Narrative language [*la langue du récit*] also possesses these dual forms of articulation and segmentation and of form and of meaning. However, Barthes confronts the inevitability of distortion in narrative, i.e. when the form of narrative is characterized by a "[distension] of signs over the length of a story and that of inserting unforeseeable expansions into these distortions" (Barthes 1977: 117). Distension indicates when the logical linearity of the narrative is disturbed (predicate before subject, for example).

While Barthes did not articulate a necessary relationship between functional units and codes, they must be assessed in an integrative manner to construct a systematic apparatus by which to perform narrative analysis in a scientific context. Connotation refers to plurality, the infinite meanings generated by a text. Functions correspond to the levels of units that engender that meaning, intrinsically plural in nature. That meaning can be interpreted through a linear trajectory (integration of units), or a distortion (Barthes terms this "dystaxia") of that linearity. These movements of linearity and distortion organize the narrative into discontinuous and continuous elements. Narrative thus appears as a succession of tightly interlocking mediate and intermediate elements; dystaxia determines a "horizontal" reading, while integration superimposes a "vertical" reading (Barthes 1977: 122). In other words, reading is not simply reading from one word to the next, but reading from one level to the next: To understand a narrative is not merely to follow the unfolding of a story, it also to recognize the construction in "storeys", to project the horizontal concatenations of the narrative "thread" on to an implicitly vertical axis (Barthes 1977: 87). In other words, the thematic turns of the story must be seen from a horizontal set of relations as well as perceiving meaning as operating from a vertical axis. Barthes uses Edgar Allen Poe's short story "The Purloined Letter" to illustrate his articulation of the horizontal and vertical axis of narrative[1]:

> Perhaps I may be allowed to offer a kind of apologue in this connection. In "The Purloined Letter," Poe gives an acute analysis of the failure of the chief commissioner of the Paris police, powerless to find his letter. His investigations, says Poe, were perfect "within the sphere of his speciality"; he searched everywhere, saturated entirely the level of the "police search", but in order to find the letter, protected by its conspicuousness, it was necessary to shift to another level, to substitute the concealer's principle of relevance for that of the policeman. Similarly, the "search" carried out over a horizontal set of narrative relations may well be as thorough as possible but must still, to be effective, operate "vertically": meaning is not "at the end" of the narrative, it runs across it; just as conspicuous as the purloined letter, meaning eludes all unilateral investigation (Barthes 1977: 87).

Therefore, the horizontal and vertical axis generate meaning by their relation to each other: they allow the reader to shift from one level of movement of narrative trajectory to another, analogous to the change in method that takes place vis-à-vis the search for the letter. To extend this analogy, Barthes asserts that as the police did not find the letter, meaning would escape the reader through a simply horizontal reading; in turn, it is necessary to discern meaning that ostensibly runs through the story on a vertical level, much like finding the letter in an obvious place.

In addition to the functional level of narrative there are the action and narration levels (as mentioned earlier as being the two remaining organizational levels of narrative). Narration is an intricate analysis of narrative as communication between receiver and sender, while the level consists of the structural status of characters and agency of actors (Barthes 1977: 109, 104).

---

[1] "The Purloined Letter" is about a queen who has stolen from her an incriminating letter by a Minister, who blackmails her. She goes to the police to retrieve the letter; the police, to no avail, undergo an exhaustive search in the Minister's apartment to find the letter. However, one of the characters in the story succeeds in finding the letter by simply looking for it in an ostensible place, which ironically, is placed in plain view to all.

Barthes' method of narrative analysis offers a model or paradigm for assessing other scientific forms of communication. In Chapter 3 of this book, the method set up in this analysis forms a model for dissecting the trajectory of scientific statements in terms of its units and demonstrating, at least for the statements in question, their plurality. This implies that accompanying the denotative qualities of each scientific enunciation, there are connotations according to how Barthes set up the concept of connotations. The scientific method is reformulated in the context of the narrative levels of organization of function, action, and narration (discourse) in a demonstration of how the theoretical movements of the method can very much be compared to the properties of storytelling. I also state that this has implications for experimental molecular science and how objects of analysis arise in the laboratory.

I am using the scientific statement as my object of analysis and ascribe its denotative and connotative levels of meaning through narrative discourse in Chapter 3. Barthes' insights into how narrative is structured can be transitioned into the structural levels of meaning in a scientific enunciation. As these statements are assigned the categories of function, actant and discourse, ambiguity or open-endedness is seen as the result of narrative analysis, a main objective of narrative inquiry.

Greimas: Actants, Isotopies and the Semiotic Square

Much of my analysis in Chapter 4 deals with a semiotic dissection of a scientific paper in the context of the work of A.-J. Greimas, the major French semiotician of his era. In this section, I will describe semiotics and its relevance in this context, and explain how specifically I am using the concepts of semiotics. For this type of explication, it is necessary to turn to Greimas' semiotic project. Greimas differed from Barthes in his narrative analysis in his new approaches with such terms as 'narrative units,' 'the action code,' continuities and discontinuities, and his comparably different ways of explicating how meaning is arrived at in narrative.

The word "semiotics" comes from the Greek root *seme*, as in *semeiotikos,* an interpreter of signs. Semiotics as a discipline is simply the analysis of signs or the study of the functioning of sign systems. In a sign system, one side of the sign is called a signifier. A signifier is the thoroughly material aspect of a sign: it is a verbal "sound image."

Inseparable from the signifier in any sign—and indeed, engendered by the signifier—is the signified. This is a mental concept; if we take the word "dog" in

English (made up of the signifiers /d/, /o/, and /g/), what is engendered for the listener is not the real dog but a mental concept of "dogness."

There are two aspects of the relationship between signifier and signified: they are inseparable (like the two sides of a piece of paper) and arbitrary. For example in different languages there are different signifiers for the English word "dog" (the French has *chien*, while the German has *hund*). That is to say, there is no natural reason why the signifier "dog" should engender the signified. The connection between the two is arbitrary. The only reason that the signifier does entail the signified is because there is a conventional relationship at play. A sign signifies by virtue of its difference from other signs, and it is this difference which gives rise to the possibility of a speech community.

Greimas appropriates this sign-signifier-signified system in articulating a narrative theory:

> Given this, one can see the development of a theory of narrativity that would justify and legitimize the analysis of narrative as a methodologically self-sufficient area of research consists not only in the refinement and formalization of the narrative models obtained through increasingly numerous and varied descriptions, but also, and above all, in the inclusion of narrative structures as an autonomous instance within the general economy of semiotics, considered to be the science of signification (Greimas 1983: 65).

Greimas sets up a model of narrative based on semiotics, predicated on the differences inherent in the sign system. The semiotic square, described below, is one of his major contributions to literary theory and is based on a system of contradictions in narrative units. He develops the concepts of isotopy and actants and elaborates on the levels of narrative structures, which he analyzes what he terms syntagmatic and semantic components.

In Chapter 4, I merge the trajectories of this semiotic analysis and an example of a scientific text by determining how meaning is generated through highlighting the syntactical and semantic properties of the text. This entails performing the Greimasian act of extracting isotopies, i.e., distilling scientific terms that engender their differential opposite from the essential argument derived from the paper.

Isotopy is a term borrowed from physical chemistry and translated for semantic analysis in the context of its new field of application (Greimas 1983: 160). The isotopy generates the four terms of the semiotic square:

> In its operational character, the concept of isotopy first of all designates iterativity throughout a syntagmatic chain of classemes which assure a discursive utterance its

> homogeneity. From this definition, it is clear that the syntagm joining at least two semic figures can be considered the minimal context allowing the establishment of an isotopy. Thus regarding the semic category which subsumes two contradictory terms: the four terms of the semiotic square ... would be called isotopies (Greimas 1983: 160).

Isotopy is the principle of equivalence of unequal units; it allows the semantic analysis of discourse beyond the confines of the sentence (Schleifer 1987: xxvii). Isotopy is the conceptual principle that holds together precisely the disparate elements of any utterance or statement. (Within a scientific problematic generated in a publication, isotopies (natural negative units) are systematically set up as a property of the scientific argument that the researchers would like to assert, which I argue in Chapter 4). Isotopies designate iterativity, i.e., that one set of units in a text has a necessary relationship to another set. I argue that this strategy highlights how relationships are realized between scientific objects.

Greimas's narrative analysis is based on the elementary structure of signification. The elementary structure is a structure of binary opposition without positive terms, what Greimas calls binary semic categories. Semes are minimal units of content. For example, in the sememes S and non S, non S is not the absence of S, but its "contrary." There are three fundamental semantic relationships that relate to how the semiotic square is formed: (1) the relationship of conjunction/disjunction, which is a binary relationship; (2) hyponymic/hyperonimic relationships between the parts and the whole, "between each of the semes taken individually, and by the entire semic category;" and (3) hypotactic/hyperotactic relationships between elements of different categories (Greimas 1983: 12). The hypotactic/hyperotactic relationship is a structural method to describe the relation between a word and its literal and figural meanings. For example, the word 'head' could be hypertactic, referring to the part of the body covered by skin and hair or only the bony part, the skull; or hypotactic could refer to a discrete organism, such as a head of a company.' It is the context of semic variables that account for effects of meaning in these relationships. In my analysis, I consider the concept of nuclease digestion to be in complex hypotactic relation to the chromatin activity since the term has multiple determinants according to its context.

Therefore let us say we have four terms: S(1), S(2), S(1a) and S(2a). The relationships between them that account for these relationships would be set up such that S(2) is the binary opposite (or contrary or contradiction) to S(1), that is non S(1), while S(1a) is the negation or antinomy of S(1), that is, –S(1). This translates into the semiotic square:

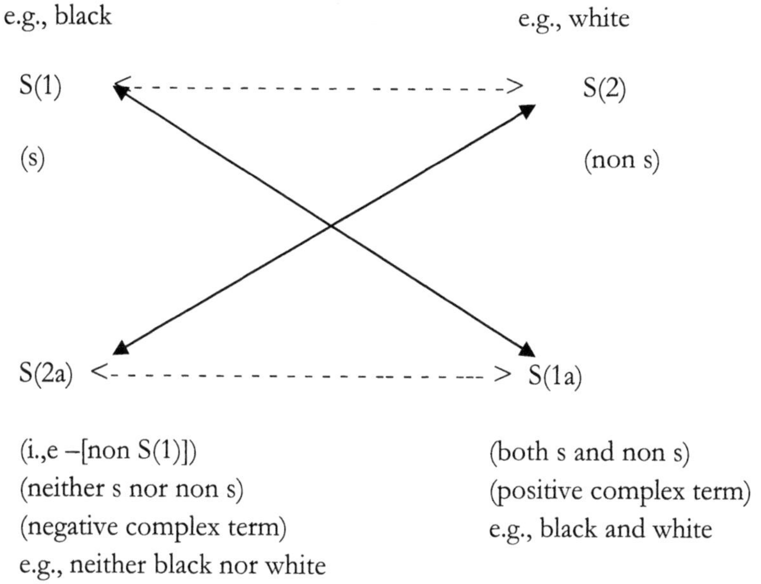

Figure 1 The Semiotic Square

The semiotic square is a logical map of structural possibilities: for any content which can be understood as itself analyzable into binary oppositions (S vs. non S), the square, repeated and superposed, will exhaust the logical structural relations between its minimal elements (Schleifer 1987: xxxiii).

The basic narrative utterance takes the form of NU (narrative utterance) = F(A) (function of the actant), in which doing, being a potentiality of the process, is designated actant (A) (Greimas 1987: 71). Narrative utterances are syntactic utterances, that is, they are independent of the content with which any given doing can be invested. Elements of the F (function) and A (actant) utterances are isotopes: Any semantic restrictions of F will have necessary repercussions on A, and vice versa. Greimas offers an analogy: the actant is an isotope of its function in the same way as the agent is the isotope of its verb (e.g. 'fisherman' and 'to fish'). Syntactic utterances convert into narrative doing the fundamental semantics that constitute the message of narrative. Meaningful discourse is articulated in utterances.

In Greimas' theory of narrativity, semiotic and narrative structures have two components: syntactic and semantic, which possess deep (fundamental syntax and fundamental semantics) and surface (surface narrative syntax and narrative semantics) levels. An example of a semantic structure is the semiotic square, an organization of semic units in which isotopies designate contradiction. Syntactic operations, as Greimas terms them, orient the narrative in sequential, segmented ways. They are oriented and thus "predictable and calculable" (Greimas 1983: 70). They can be compared to Barthes' cardinal functions.

In this context, Greimas evokes the concept of "conversion," which I refer to in my semiotic investigation of a scientific paper. Conversion thus rests on the equivalence between the syntactic operation and syntactic doing on the one hand, and between syntactic doing and elementary utterance of the actantial doing on the other (Perron 2000: xxxi). Conversion also allows for the semantic operations of the text to operate on the discursive level. For example, in "Chromatin Structure and Cardiac Gene Expression," the review article I perform a semiotic analysis upon, nucleosomal positioning is a semantic operation that correlates gene expression and chromatin activity. The semantic level leads to the juxtaposition of figural elements in the text that incorporate the thematic elements of the paper's argument (see Chapter 4). The processes of figurativization and thematization operate at the discursive level of the narrative.

In addition to semionarrative structures, there are discursive structures, composed also of a discursive syntax and a discursive semantics. Discursive semantics consist of thematic and figurative structures and operate on the levels of a scientific paper, with the thematization of the biological terms represented on a figurative plane. (I demonstrate how the scientific objects of analysis are discussed in terms of their figural representation that organizes the thematic trajectory of the paper, "Chromatin Structure and Cardiac Gene Expression"). It is at the levels of narrative and discursive analysis that allow for meaningful utterances:

> The identification of two autonomous yet linked levels—narrative and discursive—accounts quite well for the ambiguous manner in which the subject of the narration functions. Both the narrative program established by a discursive configuration in which a figure that is only barely posited proposes a relatively constricting figurative sequence. The discursive manifestation of narrativity is no more than an integration of its *semantic component* into the narrative objects generated by narrative grammar. This semantic component is presented in its syntagmatic form and it is already developed as the form, not the substance, of the content. The conjunction of these two instances—narrative and discursive—brings about the investment of contents within the canonical grammatical forms of narration and allows for the production of meaningful narrative messages (Greimas 1983: 116).

This passage best summarizes how narrative operates at its distinct levels: that is, through the essential process of integrating the semantic operations at the narrative level into its discursive components. This highlights the trajectory of conjunction; it is the interconnection or convergence of discursive and narrative levels that bring about meaningful utterances.

In Chapter 4, I introduce the concepts of modal value and narrative competence. Modal values are properties of modal utterances that represent knowing-how-to-do and being-able-to-do on the part of the subject. At the narrative level, we propose to define competence as the wanting and/or being-able and/or knowing-how-to-do of the subject that presupposes performative doing, which, in vague terms, is the difficult task or test the hero is supposed to accomplish in a narrative (Greimas 1987: 109). The competent subject and the performative subject are two instances of the same actant. The subject must acquire a certain competence before it can become a performer.

Deixis constitute certain axes on the semiotic square. For the subject to become able to produce or read narrative messages, the subject first must have access to an elementary structure that articulates signification into isotopic sets for which the semiotic square can serve as a model (Greimas 1987: 108). The semiotic square produces a positive deixis, or an axis if you will, which is a combination of S(1) and S(2a), and a negative deixis, which is the addition of S(2) and S(1a). In a narrative, these deixis characterize such oppositions that are invested in good and evil contents, such as hero vs. traitor, helper vs. opponent, etc.

Greimas' theoretical framework of narrative allowed him to develop models for the social implications of his abstract conceptualizations. Greimas built sociolinguistics, the branch of linguistics that studies social connotations, and articulated it within the framework of the linguistic sign system: Social connotations are not linked to the particular articulations of the linguistic signifier; instead the arbitrary relationship exists between gaps in the signifier and the gaps of the social signified they engender. Linguistic phenomena of all sorts and linguistic units of all dimensions, situated at all levels of language, can be socially connoted. Hence we can see that the very absence of the motivation of sociolinguistic connotations prevents us from undertaking the study of such connotations starting from a single signifier, that is, from the description of natural language. Only scientific praxis can validate sociolinguistic connotations (Greimas 1987: 203).

Greimas' narrative framework offers powerful tools for assessing the objects of scientific inquiry, in this case discovery in molecular biology. The semiotic square, based on isotopies, sets up relations of signification among scientific terms and determines the epistemological process upon which scientific observation is constructed. As I argue in Chapter 4, the rhetoric of scientific communication is based upon narrative properties of a review article dissected semiotically.

In this context, Ian Hacking describes the concept of "meaning-incommensurability," that stems from asking about the meaning of terms that stand for theoretical, unobservable entities. How do names for theoretical entities or processes get their meaning? Theoretical terms refer—almost by definition—to what cannot be observed. How do they get their meaning? (Hacking 1983: 72). In this context, Hacking elaborates: "We can give some meanings by definition. But in the case of deep theories, any definition would itself involve other theoretical terms. Moreover, we seldom use definitions for starting an understanding. We explain theoretical terms by talking theory. This has long been suggested that the sense of the terms is given by a string of words from the theory itself. The meaning of individual terms in the theory is given by their position within the structure of the entire theory" (Hacking 1983: 72). Hacking then states that "mass" in Newtonian mechanics has a different meaning than "mass" in relativistic physics, as the term "planet" has different meanings in Copernican and Ptolemiac conceptions.

Hacking articulates a scenario where there is the sentence that one theory predicts and the other denies. Narrative theory adds an additional perspective to this epistemological quandary, for confrontation and struggle between the central components are crucial feature in the determinability of an object in narrative. According to the Greimasian paradigm, the subject and anti-subject are contradictory entities in a negative relationship that resolves itself through the function of dominance. The terms "mass" and "planet" have incommensurable meanings based on which theory is applied to define them. Scientific terms, as I discussed earlier, generate their signification by their agonist relation to their binaries. In other words, we could not distinguish the distinctions between terms derived from Einsteinian physics and those from Newton's theories except by their enunciation through the differential nature of language. This theoretical impasse can be addressed through the conflictual relations among homologous scientific terms. "Mass" has a different configuration in Newtonian mechanics versus relativistic physics, and this distinction indicates an agonist struggle between the two paradigms. Thus, the term "mass" self-reflexively evokes its binary; and, this meaning can be evoked through a semiotic framework that insists on the determinability of the term. More precisely, the definition of the term mass (or planet), emerges from the theories that produce conflicting predictions, and argues for the differential nature of language to produce meaning. This highlights how the semiotic method can produce a construct for assessing epistemological conflict between scientific terms that occurs because of differing scientific paradigms.

Putting it all Together: Barthes and Greimas in the Context of Molecular Biology

The analysis that I have tried to put together is based on a formal (logical) analysis of science. This is predicated on two assumptions: that science has elements of narrative that can be formally analyzed and that the practices of laboratory investigation in genetic transcriptional and chromosomal research can be constituted as fundamentally narrative in nature. The causative and temporal events of narrative can be assimilated to the properties of scientific statements and the conceptual perspectives they give rise to. Chapter 3 treats an area of controversy in molecular biology that, I argue, can be resolved by narrative analysis based on Barthes' and Greimas' apparatus.

Chapter 4 deals with a semio-narrative analysis of a scientific publication, "Chromatin Structure and Cardiac Gene Expression." The scientific terms at stake in this paper, as I illustrate, operate as isotopies that can be mapped out in a semiotic square. The binaries that the paper sets up suggest that the biological phenomena the authors argue for can be resolved on a semio-narrative level and discursive level, attributing semantic and syntactic properties to such terms as *nucleosomal perturbation*, *DNaseI hypersensitivity site*, and *active* and *passive chromatin.* I try to ascertain the epistemological significance of this type of analysis for the scientific process and method.

Both analyses have implications for molecular biology that can be assessed legitimately on epistemological and structural levels with discrete logical aims. I link my objectives to the aspects of narrative paradigms and models, and current theoretical formulations of narrative theory. My purpose in doing so is not to conflate all of narratology, but to evince that science, when put in such theoretical contexts, is transformed into a cultural construct possessing narrative. In short, molecular biology and genetics are considered from the perspectives of narrative theory and of semiotics.

Chapter 3

Reassessing Molecular Biology in the Context of Advancing a Narrative Analysis

Any type of endeavor that attempts to divert the scientist's attention away from the practice of science may seem subversive to scientific goals and aims. A scientist is caught up in the hermeneutic search for truth, to discover facts that ostensibly mirror nature. This account may be simplistic; however, its tenets are those to which most scientists subscribe. Yet in regards to molecular biology, the object of my study, this account does not do service to the enormous complexity of that science, the differing methodologies in experimental molecular biology that lead to new data, the contradictions that emerge in assessing that data and the ensuing debates that arise out of conflicting claims over scientific information. This ostensibly serves as a springboard for analyzing the workings of science and the scientific process through an understanding of the workings of narrative.

I establish the assumption that molecular biology has a narration and epistemology behind it. My rationale consists of the presumption that molecular and cell biology fall under the rubric of the natural sciences and lay collective claim to being considered "sciences."[1]

As I will argue, it is the rigorous methods, models and approaches that have been developed in molecular experimentation that allow for an analysis via the tools of literary theory to reveal the underlying narrative structures and epistemology behind this branch of the life sciences. Cognizance of these rhetorical moves would make for more precise analytical discovery, greater organization in science and contribute to greater methodological self-awareness among molecular biologists. As an extension of my introduction, I would like to articulate an epistemological justification for the intellectual relevance of narrative theory to the analysis of scientific language and methodology inherent in the laboratory output of scientific investigations, particularly in the natural sciences.

Carl Hempel discussed what may be considered one of the most famous articulations of the philosophy of the natural sciences. "Deductive-nomological

---

[1] E.D. Klemke comments more precisely on this distinction. Medicine is considered an applied science which must be distinguished from the pure sciences, such as molecular biology: "The findings of the pure, empirical sciences are applied to disciplines which fulfill various social, human purposes, such as building houses or roads or health care ...The natural sciences are primarily law finding, whereas the social sciences [such as psychology and sociology] are fact finding" (Klemke, 30).

explanations satisfy the requirement of explanatory relevance in the strongest possible sense: the explanatory information they provide implies the explanandum sentence deductively and thus offers logically conclusive grounds why the explanandum phenomenon is expected" (Hempel 1966: 52).

Hempel continues that "corresponding general laws are always presupposed by an explanatory statement to the effect that a particular event of a certain kind G was caused by an event of another kind, F...whenever an event of kind F occurs, it is accompanied by an event of kind G" (Hempel 1966: 53). This effectively enunciated the logical positivist model of science, which was a model for explanation in science, but discounted the veracity of theoretical entites (Hacking 1983: 48).

In this context, I place science within the crucial confines of the experimental method it offers as the basis for scientific discovery. Here, experimental fact is the object of inquiry. Rather than emphasizing its discursive properties, I would like to conceive of science as an open-ended project, whose narrative properties lead to the temporal perpetuation of the scientific process as a result of its inherent ambiguity. As molecular biologists make discoveries in the form of scientific enunciations, narratologists, in turn, can systematize these discoveries.

I use David Gooding's formulation of experimental maps as a potential comparison with my narrative analysis. Gooding uses the concept of experimental maps, in which experiment is represented as a process, to visualize the complexities of experiment and that the changing relationships between experimental practice and argumentation emerge from a comparison of these maps and helps us understand how scientists move from local, situated uses of particular practices to the realm of general, theoretically significant argument (Gooding 1992: 67). I attempt to attribute the same objective to a narrative analysis of scientific communication, which could constitute a type of experimental map, and address the originality of this type of thematic framework.

Gooding's experimental map is a reconstruction of the forms of narrative that occur during the various stages of scientific discovery. Notebooks enable representation, communication, argument; drafts of papers enable ordering, description; research papers enable communication, persuasion and dissemination (Gooding 1992: 78). Gooding characterizes each stage as occurring either in real-time, prospective and retrospective frameworks, similar to how narrative is fashioned.

In turn, I will show how the narrative methods of the French theorists illuminate this comparison.

We may begin at a crucial moment in narratology offered by Roland Barthes. In *S/Z*, exhaustively dissecting the short story "Sarrasine" by Balzac produced a series of codes that would ostensibly be evoked for other types of narratives. For Barthes, the narrative constituted a manner by which the roles of the story and the processes of the plot unfolded into certain voices: "Or again: each code is the one of the forces. According the Barthesian paradigm, each signified has hermeneutic value; every action of meaning is an action of truth (*S/Z*, 62).

Greimas, whose response to Barthes was his analysis of Maupassant's short story "Two Friends,"[2] acknowledged the power of Barthes' analytical move, but went further in his engagement of the text in positing an enterprise of semantic analysis that refined narrative operations in terms of actants and actantial roles. Maupassant, Greimas argues, can be read in the manner of reading a poem—in fact, a "symboliste poem" (Greimas 1988: 28): "the short story, as a genre," he writes "can be considered as the equivalent in prose of a poem because of its simultaneous paradigmatic and syntagmatic structure" (Greimas 1988: 12). If *S/Z*, with its plural readings, fragments, digressions, and "decomposition" is an archetypal "metonymic" reading, the reading of prose as signifying chain then *Maupassant* can well be an archetypal "metaphoric" reading, not a "chain" of signifiers, but the reading of poetic substitutions of signifieds (Schleifer 1992: 150). In *S/Z,* Barthes distinguishes between classic "readerly" texts which are "decidable, continuous, totalizable, and unified into a coherent whole based on the signified" and "writerly" texts which are "infinitely plural and open to the free play of signifiers and of difference, unconstrained by representative considerations, and transgressive of any desire for decidable, unified, totalized meaning" (Johnson 1980: 6); in *Maupassant* Greimas erases this opposition. He offers, instead, a "readerly" totalizing reading that eschews the referential, focusing on the play of signifieds without being constrained by "representative" considerations (Schleifer 1992: 151).

This unfolding of narrative as explicated by Barthes and Greimas has an inter-disciplinary parallel in its applicability to laboratory procedure and convention. The path to discovery in molecular biology is an arduous one, with repetition, troubleshooting, and perpetual experimentation the norm. A "free play of signifiers" operates at this level and the desire for a "totalized, unified"

---

[2] Alain J.-J. Cohen, personal communication.

meaning is subverted in the process of attempting to answer questions. The apparatus of the system must be optimized for the observations to be considered valid, and if so, they can be considered results. A successful result leads to discovery, which leads to more questions and to the advancing of subsequent hypotheses. In many ways, a scientific theory is a semantic construct. Lexical choice and propositional structure are instrumental to scientific investigation and serve as a framework to drive inquiry of the natural world (Frawley 1988: 335). Understanding the narrative trajectory of a scientific enunciation is to filter out its rhetorical advantage (Porter-Abbott 2002: 146). The rhetoric of a scientific text is the object of discussion here, for it is at the pivotal moment of persuasion when rhetoric comes into play. Through narrative analysis we can arrive at points where the argument of a scientific text sustains itself. This sustainability presents itself through the use of linguistic metaphors, a parameter of narrative. It is clear that a metaphor is a sign-process (Morgan 2000: 473). A classical example of the power of metaphor is the discovery of the benzene ring by Kekule, as after much pondering he dozed and saw flames turn themselves into rings, like snakes eating their tales (Morgan 2000: 474).

Metaphors, images and contradictions dominate the mass of information circulating the arena of molecular biology, genetics and molecular science. Boyd (1979: 360) has defined theory-constitutive metaphors as figures of speech where "metaphorical expressions constitute, at least for a time, an irreplaceable part of the linguistic machinery of scientific theory." As I will try to highlight the narrative science can serve as a direct framework to elucidate, clarify and characterize the laboratory practice in modern molecular biology, and it is the *method* of theoretical and literary analysis, the analytical procedures which define these approaches, that has relevance for the scientist. This method can offer an epistemological framework, or a manner of critical thinking, that science can make use of in cases where the experimental apparatus only leads to debate. Applying the methods of narrative analysis can reconcile conflicting results from different experimental approaches, further advancing the process of experimental molecular biology. The issue must be addressed in terms of how each narrative unit functions to accord levels of meaning within scientific statements. In turn, I elaborate how the scientific process is perpetuated by applying the narrative method. What the narrative method does is offer a paradigm for mobilizing the arguments presented in a scientific text, and serves as an impetus for further research. It is not simply a matter that the text produces ambiguity, but that Barthes' and Greimas' semiological assessments can situate molecular biology within a framework beyond mere theory but into the practicalities of laboratory investigation, and lead to productive formulations of new ways of conceiving of discovery. As I will show, scientists think of original perspectives to think of biological systems through a process that can be illuminated through narratology.

Each theorist that I will be referring to in this context has diverse and differing assumptions and objectives, and he must be concretely distinguished in terms of his trajectories. If Barthes posed his codes in his analysis of lexias while Greimas extracts his isotopies, we can pose a singular set of analogical constructions that delimit how the scientific process undergoes a similar instantiation of positing narrative frameworks and further elaborate on scientific statements, in the context of a reassessment of the scientific method with regards to molecular genetics.

The nature of the narrative endeavor, in light of both Barthes' and Greimas' analysis, concerns a formulation of narrative, that in essence, reformulates it, thereby producing the "writerly" text. This is the approach I take towards scientific statements, when examined through these lenses, is that they are reformulated in order to perpetuate the scientific process. What scientists hope to discover can form the basis of narratological efforts, i.e., molecular biological enunciations are the object of inquiry for narrative theory. In assessing this progressing, there are cardinal functions, necessary for the perpetuation of the scientific narrative by ascribing elements of causation to the structure of the statements in molecular biology.

Barthes, Greimas, and an Examination of a Scientific Narrative

Barthes offers a profound examination of "Sarrasine" that reconstructs the characteristics of a story to demonstrate and define the elemental functions of each signifier-coded function. He does this by evoking the interplay between connotation and denotation.

> Structurally the existence of two supposedly different systems—denotation and connotation—enables the text to operate like a game, each system referring to the other according the requirements of a certain *illusion.* Ideologically, finally, this game has the advantage of affording the classic text a certain innocence: of the two systems, denotative and connotative, one turns back on itself and indicates its own existence: the system of denotation; denotation is not the first meaning, but pretends to be so; under this illusion, it is ultimately no more that the last of the connotations ... doesn't a sentence, whatever meaning it releases... it would seem, appear to be telling us something sim-

ple, literal, primitive: something *true*, in relation to which all the rest is literature? (Italics Barthes, Barthes, *S/Z*, 9)

Denotation is the literal substance but does not compose the overarching essence of meaning. It interacts with connotation in ascribing to the literary narrative a sense of thrust, a backdrop of acknowledgment with acquiescence of the moment when the unknown is about to arrive. When one reads *S/Z*, Barthes makes clear that the castrato's (in "Sarrasine") denotative feminine essence that masks the connotative masculine operatto. What transpires in levels of hermeneutic engagement is the "primitive" with the implicit.

The denotation of a statement in science would fall in this category:

"Molecule X binds to molecule Y and activates transcription of gene Z."

To comprehend the connotation relies on implicit knowledge that molecules X and Y must traverse the nuclear membrane to bind to a histone-DNA complex, yet this is acknowledged but not proven nor fully known. The scientific narrative must be further constructed by understanding that its denotation, like literary denotation, is the simple and primitive, and yet must be underscored by an analysis of the connotative.

One of the experimental methods that lead to discovery in modern molecular biology is current-day laboratory routine, manifested in a formal scientific publication as: observation→ result indicating argument of the paper→ discussion of results→ further experimentation. Considering the Barthesian apparatus, this sequence acquires the properties of narrative elements:

Observation a(1): molecule X binds to molecule Y. These are constructs that mark the physical representation of biochemical affinity according to Michealis-Menten parameters.

Observation a(2): Molecule X binds to molecule Y and activates gene X. Two narrative trajectories are occurring here. Gene Z is a construction that forms the teleological boundaries of the bound complex of molecules X and Y. The binding of X to Y imparts the necessary contingencies for the activation of gene Z. I would argue that this is a deductive-nomological explanation that the binding of molecules X and Y *causes* the activation of gene Z.

Observation a(1) and observation a(2) are denotations that set up the logical deduction a(1) Molecules X and Y must traverse the nuclear membrane and the logical deduction a(2) Molecules X and Y in concert must disrupt the nucleosomal-DNA complex to expose the naked DNA for binding, both consisting of connotations that set up further questions to be posed and experiments to be performed, extending the biological narrative of denotation and connotation. The questions consist of whether molecule X binds to molecule Y before entering the nuclear membrane, or if both molecules traverse the nuclear membrane independently.

In essence, a discourse is being set up within the narrative properties of the biological story, and to dissect this discourse it is necessary to understand the ontological characteristics of a complex, laboratory-derived statement such as "Molecule X binds Molecule Y and thereby activates gene Z." Despite the enormously complicated biology, narrative theory provides an analysis to address the conceptual framework of such a result in terms of assessing the representational power and equivocal or unequivocal nature of such a statement. In essence, the potential ambiguity of the statement is arrived at from narratological methods in prototypical fashion. Further, novel ways of thinking of this system emerge through narratology. As I have described, if molecule X functions to bind to molecule Y, then the teleological effect of the transcription of gene Z remains. Also remaining is the focalization resulting from narratological analysis that transcription is a spatio-temporal phenomenon effectuated by the allosteric binding of molecules. By privileging this molecular biological model, transcription is metaphorically represented in a specific manner.

Furthermore, consider Barthes' statement on character and figure: "When identical semes traverse the same proper name several times and appear to settle upon it, a character is created. The character is a product of combinations: the combination is relatively stable (denoted by a recurrence of semes) and more or less complex (involving more or less congruent, more or less contradictory figures)" (*S/Z*, 67).

Here, the seme of the binding of molecule X binding to molecule Y creates an insolvable character in that the activation of gene Z is an implicit exigency of the affinity between these two molecules. This brings up reflective positions on how to think of the phenomena: Is the semic combination of molecule X and molecule Y independent of their subsequent activation of gene Z or is it plausible for molecule X and Y to bind *ipso facto* without exerting their effects on gene Z? This brings up questions that have implications for modern molecular biology since to pursue the experimental consequences of the logical deductions from their connotative value would necessitate a system that is predicated on addressing the teleological assumption of this scientific statement.

Teleology is privileged by narratological analysis in terms of its predication on spatio-temporal causation, and each derivation from an analysis of the scientific method through narrative theory extends the association between the objectives of molecular biologists and narratologists. Roland Barthes once made a similar claim, describing narrative as a culturally sanctioned application of the fallacy *post hoc, propter hoc* ("after this, therefore because of this"). To paraphrase, narrative understanding depends fundamentally on a generalized heuristic according to which interpreters assume that if Y is mentioned after X in a story, then X not only precedes but causes Y (Herman 1999: 176). The heuristic in the claim that molecules X and Y activate gene Z in unison is that molecules X and Y possess causative agency in their activation of the transcriptional apparatus. This is predicated upon the activation of gene Z.

One can set up a model and construct its epistemological relationship to the investigative apparatus as it is played in the modern molecular biological laboratory. As in a literary model, the narrative must compose levels of meanings within its linguistic parameters (Barthes, *Image Music Text*, 88). In the following passage, Barthes elucidates his manner of how narrative work is organized:

> It is proposed to distinguish three levels of description in narrative work: the levels of "functions" (in the sense this word has in Propp and Bremond), the level of "actions" (in the sense this word has in Greimas when he talks of characters as actants)[3] and the level of "narration" (which is roughly the level of "discourse" in Todorov). These three levels are bound together according to a mode of progressive integration: a function only has meaning insofar as it occupies a place in the general action of actant, and this action in turn receives the final meaning from the fact that is narrated, entrusted to a discourse which possesses its own code (Barthes 1977: 88).

Function, action and narration, and its elemental derivative discourse, can be formulated in a semantic model of scientific methodology that meets Barthes' requirements of narrative. Consider the following scientific enunciation:

(2) Fluorescence in-situ hybridization (FISH) reveals variation in chromosome number (aneuploidy) in cancer cell lines, thereby generating chro-

[3]Barthes here is referring to Greimas' *Structural Semantics* rather than Greimas' *Maupassant*.

mosome instability and interfering with mitosis.

This statement lies in the province of research into genetic instability. One can discern a similar epistemological framework to statement 1 that further indicates the narrative structure of the experimental method in chromosomal biology:

Observation a(1): the technique of FISH, whereby chromosomes are stained and viewed under a fluorescence microscope, demonstrates that there is a greater or lesser number than the 46 homologous chromosomes.

Observation a(2): The preceding observation is contingent upon viewing chromosomes garnered from tumorigenic cell lines, or lines with abnormal cell growth.

Observation a(3): Observation a(1) directly implies that this aneuploidy is directly related to the mitotic cycle, or to how chromosomes replicate.

Aneuploidy and mitotic interference are actants that serve as functional units in their agonist role in cellular aberration. In other words, aneuploidy and mitotic interference operate as rhetorical players on the discursive level of semantics. This correlative conclusion underscores that the direct relation between aneuploidy and the mitotic cell cycle further progresses the narrative trajectory of the scientific experimentation since an analysis of scientific statement (2) does not reveal which occurs first: the mitotic cell cycle displacement or the appearance of aneuploidy characteristics of the chromosomes in cancer cell lines. This corollary of the discourse of the system necessarily corroborates the argument of the paper.

This highlights a narratological model that would be formulated as such in terms of the hypothetico-deductive method:

Characterization (Function 1)→Hypothesis→ Prediction→ Observations (Actants 1)

↓

Articulation of Experiments and Implications (Discourse)::

⇕

Observations (Actants 2) → Ambiguity (Unit of Meaning unresolved from Results)→ Characterization (Function 2)→Hypothesis→ Prediction→ Experiments

Figure 2 A narrative configuration of a category of experimental method

The scientist has a position in this model using the experimental rationale played out in the laboratory. Benchwork remains the hallmark of this narrative role and publication is the manifestation of this discourse. Ultimately, the scientist's productivity foregrounds the qualities of narrative as evinced in a pragmatic modification of the discovery in molecular biology through the analytical movements offered by narrative science. The complexity of a narrative can be compared to an organizational profile chart, that is capable of integrating backwards and forwards movements; or more accurately, the chart integrates in various forms which compensates for the seemingly unmasterable complexity of units on a particular level (Barthes 1977: 122). However, Barthes' point must be qualified, for it is the modes of segmentation of units that make the units masterable. In essence, scientific output is intimately associated with narrative production. The resulting metaphorical model that is produced is in accord with its spatio-temporal properties, revealing the narrative nature of perceiving scientific processes' ramifications, which are spatio-temporal causation, plurality and open-endedness, perpetuation and the production of differential frameworks of conceiving of systems in transcriptional regulation and chromosomal aberration.

The preceding paragraph amounts to what is a logical analysis of formal (scientific) statements on a semantic, or discursive, level predicated on the paradigms of narrative. In this light, *Maupassant*, Greimas' narrative analysis, begins with methods of segmentation; Greimas divides Maupassant's story into twelve "sequences" and further subdivides these sequences themselves into what he calls "segments," sometimes "narrative utterances" (*Maupassant*, 69). In his conclusion, Greimas notes that the book attempts to vary its methodological approaches to the text as much as possible (M: 263), and in different chapters its presents "multiplication of criteria for segmentation." Thus he uses "spatio-temporal" isotopies (M: 19), different sentences (M: 23f), repetition of sentence fragments (M:93), repetition of particular words in the text (*s'arrêter, Mais*) (M: 84, 135) or particular morphemes (*re*vint, *re*mit) (M: 253), repetition of 'interpretative action' of the characters (M: 119), logical and parallel relationships (M: 144), "formal" oppositions (M: 175), repetition of physical action (M: 191, 216) and in one case the delimiting of surrounding sequences (M: 240) (Schleifer 1992: 148). The segmentation of "Deux amis" is governed by methodological choices that assume both the 'wholeness' of signification and the efficacy of the logic of analysis. Segmentation, then, can take place on either the (discursive) level of actors or the manifestation of the (semio-narrative) level of actants. Segmentation:

> consists in beginning with actors who are manifested as discursive subjects endowed with predicates and attaining actants which can be integrated within the narrative utterances. Thus the procedure seeks a narrative organization subjacent to the discursive manifestation which will allow the articulation of the textual surface to be accounted for. In this way the behavior of discursive actors can serve the segmentation of the text, while the presence and absence, the appearance and disappearance, of the actors and the signifying variations in their predicates can be considered as demarcators of the text equal to spatio-temporal criteria (*Maupassant*, 67).

The paradigm of segmentation is demonstrated in how narrative analysis, in the Greimasian context, functions to articulate demarcations in scientific 'utterances.' Within the logical predicates of scientific statements (1) and (2) lie discursive properties that highlight the formal properties of the observations vis-à-vis the scientific method that further instantiate the narrative trajectory of laboratory experimentation. Each statement contains actants that are segmented to reveal the equivocal nature of the results that refine how one conceives of the scientific method, within 'spatio-temporal criteria' that scientific actants operate.

In offering another application of the narrative model, Porter-Abbott uses the method to explain the theory of evolution and how it attempted to counter creationism through this model. He addresses evolution in terms of complex causation, or narrative by inference and indirection. Whether the concept of evolution is right or wrong, it can be narrativized since it involves events that are explicable in terms of agential force, that is, a narrative entity. If the entire process can be described by narrative, it is narrative by inference:

> Level 1: Species → Species → Species → Species
> Level 2: Individuals dying without issue or surviving and procreating

Porter-Abbott's synthesis addresses narrative in terms of its rhetorical advantage, that is, narrativizing a scientific theory lends itself to argument. My own analysis is posited along similar lines, however, I attribute the narrative properties of the discovery and endeavor to its open-endedness. Connotative plurality is an essential component of the scientific method, and the endeavor of molecular biology is perpetuated from this framework in which scientific statements are narratively dissected. While narration has been attributed to certain scientific theories, such as the theory of evolution, this is the first analysis posited on the notion that the perpetuation of the scientific process results from its inherent narratological properties.

However, assessing scientific theory as narrative by inference, as Porter-Abbott does, leads to a crucial leap in the theoretical direction that narratology takes us in regards to biology. While narrative is concerned with the representation of events, and evolution can be characterized sufficiently as an event concerning the speciation of life, I would like to address how interpreting the connotative value of scientific statements leads to differential configurations of results at the observational level.

As I have demonstrated, these goals, when transposed, illuminate the scientific endeavor and exemplify the perpetuation of laboratory investigation. The analysis that I have conducted is a novel illustration of how science, when conceived through the perspective of a paradigm that ambitiously attempts to structure textual work, is a boundless endeavor that limitlessly sustains itself. The analogy to literature, or of the never-ending story, is telling.

Examining Conflict in Scientific Narratives and Their Claims to Lucidity

Scientific data often produces results that are inconclusive. Scientific claims, when compared together, result in discordant data. The object of this section is to elucidate the processes of conflict in methodological investigation in science using terms of narrative. My essential premise is that scientific claims are a product of traditional conflicts, which are essentially narrative in nature come down to assessing the rigor of experimentation behind scientific data. This rigor can be evaluated by delineating the narrative properties underlying the processes of producing, amassing, and presenting scientific data, and by addressing its epistemological relationship to structuralism.

Yet even with the necessary elaborations, this description begets the question, which was not addressed previously, that even if insight can be generated, why is it necessary for a scientist to be acquainted with semiotic narrative analysis? The argument for doing so will be the subject of the remainder of this chapter and lies within the fabric of molecular science *per se*. By examining examples of debate within molecular biology that we see that traditional methods of judging scientific claims and discoveries are not sufficient for assessing their veracity. Therefore, one must turn to the tools of literary narrative theory to improve our methods for evaluating new scientific data. In doing such, I will show how the metaphorical models utilized to address the problematizations of narratology are pertinent to the representation of scientific discovery.

One could argue that the scientific endeavor to present data and establish the veracity of scientific facts is itself a transparent activity that lays claim to lucidity. As a member of a scientific research laboratory, I am attempting to acquire a sense of the epistemological framework behind scientific discovery. I am in turn engaging in understanding the narratives behind science through semiotics and structuralism and contemporary theory. Within these dual constructs, science seemingly undergoes a transformation from abstract hypothetical arguments to concrete conclusions that are grounded in a sound theoretical foundation through the use of experimentation and resulting data. This whole process is presumptively void of obscurity and delusional activity. Yet, if this system is subjected to closer scrutiny, the scientific approach is manifestly wrought with questions of certainty and uncertainty and potentially may step over the dividing line between knowledge and belief.

Models serve as the basis of the methodologies of scientific data, particularly molecular biological data. Data serves as precursors for models; in turn, these models become incorporated into fact by presumptive data. However, this reliance on models and data is inherently selective and exclusive; ostensibly certain observations are assimilated into scientific reports that serve to validate the conclusions of the recorded data and protocols. Furthermore, contradictory observations and results often emerge within the mass of scientific articles. Even if assessed in light of an analysis that would scientifically evaluate contrasting data, science still cannot lay claim to being a system of lucidity in clear contradistinction to delusion. The epistemological framework behind scientific investigation (molecular biology) presupposes certain simplifications/idealizations that inherently inure the experimental method in modern molecular biology to an inevitable evidentiary crisis. These simplifications and idealizations lie in the assumptions upon which scientific investigation is conducted. As an analysis of the scientific apparatus behind how materials and methods are constructed reveals, scientific discovery results in a disruption in the reliance on proof because the basis for evidence is ultimately equivocal.

In short, while society relies on science for clear-cut answers, ambiguity and equivocation rather than conclusory clarity are the norms. In assessing the validity of any scientific assertion, its claims have traditionally been set against the standard of doxa, or opinion. In fact, the technical presuppositions of any type of scientific output are considered knowledge in contrast to belief. Contingent upon this assumption is that science is a true mirror of nature, corresponding to nature, and accurately reflecting the natural reality around us. This supposition is so strongly embedded in the structures of scientific knowledge and thought that even Michel Foucault believed the natural and physical sciences were out of the purview of his genealogical and archaeological analysis since they crossed what he termed the threshold of scientificity.

However, to question science's claims to lucidity is to subject the scientific method to epistemological scrutiny and to place scientific experimentation within a rhetorical framework. That is, one must have as the object of study the scientific methods and observations, and a mechanism to appropriate the results of this process, i.e. proof, scientific data and results, within the confines of hermeneutic structures that fall within the boundaries of truth and knowledge, and doxa and belief.

To commence with this approach involves taking as an example a scientific problem that engenders controversy within the field of molecular biology. Holding up to the light of the rigorous standards of the experimental procedure, while simultaneously undermining its claims to lucidity and objectivity through an analysis of the resulting data, is the debate surrounding the role of chromosomes in transcriptional regulation in organisms. These are keywords in molecular science but unfortunately constitute jargon to the reader unaware of the biological models upon which molecular biology is based. In an organisms, chromosomes are the bearers of genetic material, deoxyribonucleic acid. The actual DNA strand is wrapped around a core set of proteins, called histones

(Figure 17). These histones anchor the DNA in the chromosome. These histone-DNA complexes, called nucleosomes, are packaged into a tightly wrapped, condensed material called chromatin, which makes up the chromosome. Core DNA and core histones are intertwined to make a histone octamer and linker DNA is wrapped around histone H1. The histone octamer, linker DNA and histone H1 make up a nucleosome (Figure 18).

Much research has been devoted to understanding how the DNA, the genetic material, can be exposed for cellular activities. One of these activities is *transcription*, a molecular process involving many enzymes and molecules that allows for the DNA molecule to be a blueprint for an ribonucleic acid molecule, also a nucleic acid. It is this RNA molecule which serves as the template for protein production. This is the central dogma of molecular biology.

The process of transcription is initiated by the binding of specific protein molecules, called transcription factors, on specific DNA sequences called promoters. There have been several different types of promoters characterized in the last fifty years of molecular biology. One promoter, called the TATA box promoter element, is transcribed by a TATA binding protein, a transcriptional factor. The TATA binding protein, TBP and other transcription factors bind to the TATA site on DNA to form a transcription complex (Figure 19).

The central biological question asks how DNA in these tightly wrapped histone octamers is laid bare for the process of transcription. One argument states that promoters are all contained within positioned nucleosomes that are disrupted during hormone-dependent chromatin rearrangements. Of the multiple modifications required for conversion of chromatin from an inactive to active state, one consistent feature of active chromatin is the highly acetylated state of the core histones in the nucleosomes, i.e. an acetyl molecule is attached to nucleosomes so that DNA can be transcribed (Figure 20). In a recently published paper by Sewack et al., (see bibliography for complete reference) the authors showed that when certain receptors are activated, hyperacetylation of histones induces binding of the transcription factor to a specifically designated promoter (Figure 21). The paper's authors also found that binding of the TATA binding protein is intrinsically linked to hormone-dependent transcriptional activation of a promoter, and that the nucleosome severely restricts binding of the TATA binding protein to the promoter. Once acetylation occurs, TBP causes nucleosomal sliding. However, the premise of this active chromatin is that its nucleosomes are "naturally positioned," i.e. there is no mechanistic movement involved in an active chromosome that is mediated by a transcriptional state, i.e., they exist in a static state. Translational positioning of the nucleosome overlapping the TATA or promoter sequence is unchanged upon induction of a promoter (Figure 22).

The current debate surrounding this central biological question is that it was previously thought that the primary role of chromatin was to act as a passive repressor in transcription and that as it became activated in the acetylated state, it passively allowed transcription. Much evidence in assessing the role of chromatin in transcription has led to this conclusion. However, recent biochemical and genetic evidence indicates that chromatin is a dynamic structure. Upon gene activation, chromatin structure is actively changed through rearrangement of histones and DNA. This is based on the premise that transcription factors actively bind to nucleosomes since promoters are packaged within these same nucleosomes. In order for transcription binding proteins (trans-acting factors) to gain access to these promoters (cis acting sequences), nucleosomes must be perturbed. It was then proposed that transcription-inducing nuclear receptors recruit histone hyperacetylation, (which I discussed in Chapter 1) to modify nucleosomal structure. According to this model, only after nucleosomal DNA is disrupted and mobilized, are transcription factors allowed to bind to its site over nucleosomal DNA (Figure 23). Transcription factors induce relaxation of histone-DNA interaction. Nucleosomes, by dynamically moving, allow all transcription factors to bind to their binding sites. In the adult version of the gene, the formation of compacted nucleosome structure excludes the transcription factors from the gene promoter (Figure 24).

However, according to Sewack, et al, nucleosomal entities remain stably positioned even upon transcriptional induction of the gene.

So what is the answer to the biological question of whether chromosomes remain static or mobile during the process of transcription, and how can contradictory scientific claims be assessed?

I have spent some time describing the biology of this problem for the purpose of demonstrating that the juxtaposition of varying scientific claims can produce, in some cases, an epistemological impasse and may potentially undermine science's claims to lucidity and to its being an apparatus of producing objective knowledge. How can this controversy be resolved if science itself produces contrasting and contradictory positions, and how can we generate insight through research and experimental approaches of molecular biology if it engenders controversy and epistemological conflict instead of unambiguous truth? If observations, results and data mechanistically produce divergent conclusions and disparate output, how are we to assess the narrative formulation of procedures routinely surrounding research in transcriptional regulation?

I would like to posit that the nature of the conflicting claims of these biological observations lies in assessing the narrative nature of the scientific process and how it affects the quality of scientific output.

If we set out with this premise, then we may have to begin with the assumption that the preconditions to producing truth claims and incurring objective knowledges that mirror nature are not inherent within the natural sciences *per se*. We must relinquish a foundational manifestation of science; can the debate about the truth claim capacity of science be placed in the arena of the social sciences or in what Michel Foucault terms the human sciences, all as a result of questioning science's claim to lucidity?

After understanding the molecular nature of a certain debate in molecular biology and genetics, the processes of scientific investigation may be characterized as ambiguous, controversial and producing conflict.

Thus, we must ask in this context about the pertinence of reflective narrative analysis to the intricacies of science. To answer this would be to don the cap of such a theorist and ask if the ambiguity lies in the language used to describe these models of transcriptional behavior. Inherent in this capacity to organize are the implications to assess codified systems of thought and representations in ways that go beyond merely typifying them as progressive, natural or self-evident. In this context, we assess two differing forms of experimental data, passive transcription or active chromatin, for which a scientist would claim that only experiments would mandate choosing one interpretation over another.

Yet, in assessing this debate, we can state that it is the narrative paradigm that drives the scientific processes forward, under the auspices of the scientific method. For example, in assessing the debate over naturally-positioned nucleosomes or active chromatin components, we can first state that the former argument maintains that nucleosomes do not move, while the latter maintains that nucleosomes are "perturbed" and the histone octamers break apart upon transcriptional activation. Second, the differing results present their experimental apparatus and models to elucidate and substantiate their claims. Thirdly, we can presume that further experimental data, such as more accurate gel electrophoresis apparatus or use of cell lines that are more potent, potentially offer insight into this problem.

However, if we were to address the problem conceptually, or in more scientific terminology, methodologically, we can see that these two chromosomal states or "worlds" are both fundamentally justified, and their significance lies not only in their scientific claims but in their metaphorical status, and metaphor lies at the heart of narrative. Whether nucleosomes are static or mobile during the process of transcription translates into certain properties of our genetic material. In other words, the geneticist would state that if nucleosomes are static, then our genetic material is unfluctuating and immobile; and, if they are mobile, our genetic material is in a state of flux.

Narrative analysis gets at the heart of the issue because it offers a precise image/reflection of the biological moves in the laboratory, in terms of theories that are an account of what happens in the lab. It offers a metatheoretical apparatus whose object of analysis are the conflicting trajectories of scientific investigation. To evince how this apparatus highlights the scientific endeavor (in the context of chromosomal movement or lack of movement) that I have laid out, consider that a formulation of the premises underlying scientific argumentation demonstrates that this debate is founded on narrative *sui generis*. Its predicates in the context of the experimental method in molecular biology are highlighted by functional units that characterize the questions and promote a hypothesis that would potentially answer these questions. This would promote a more precise self-awareness among the activities of molecular biologists by making them more cognizant of the narrative nature of the biological movements in the laboratory.

Observation (1) Nucleosomes are not perturbed in transcript-tional regulation.
Observation (2): Nucleosomes are perturbed in transcriptional regulation.
Observation (1) contradicts
Observation (2); progression of the narrative.

↓

Characterization (Function)→ Hypothesis (a1)

The perpetuation of the scientific project through narrative analysis

Hypothesis a(1) can be segmented in the following manner: one can choose between observations (1) or (2) or articulate a medium between them, revealing a new set of actantial roles among biological actors. Observation (3) results with new discourse in the articulation of the results in published papers. The important point is that there are two focalizations that result from the explanation of one biological phenomena, in this instance, the activity of transcription. In semiotic terms, the static nucleosomes form the subject versus the anti-subject in terms of passive and dynamic chromatin. The determinability of the molecular objects, depicted as spatio-temporal metaphorical models, is performed through the linguistic value of difference. These differential categories, in which the scientific terms are set up as binaries, account for one molecular biological event and are pivotal in the formation of the divergent conceptualizations, and causation is key in this evaluation. This is crucial in the scientific endeavor (I explore this concept further in the following chapter).

The statements of transcriptional regulation and chromosomal aberration lead to equivocal interpretations through a dissection of their narrative properties. These differing perspectives emerge from the plurality of these statements that inevitably and epistemologically affect the way of looking at these phenomena. Because this constitutes a fundamental operation of constructing scientific argument, we can maintain that the narrative is an essential property of scientific communication.

How can we correlate Kuhnian notions of normal science and paradigms with a narratological analysis of molecular biology? Normal science, Kuhn concurred, does not aim at novelties of fact or theory, and when successful, finds none. New and unsuspected phenomena are, however, repeatedly uncovered by scientific research, and radical new theories have again and again been invented by scientists (Kuhn 1996: 52). Crisis emerges when there is a "decisive difference" in problem-solving between transitions, in which a "profession will have changed its view of the field, its methods, and its goals (Kuhn 1996: 85). Kuhn argues that this is a "normative" function of paradigms in terms of normal science.

Narratology attributes a connotative plurality to a text, whereby heterogeneous readings emerge from a denotative statement. Narrative analysis would be somewhat similar to Kuhn's views in the sense that we might outline a framework in which the notion of normal science is conceived as arising from a plural reading of molecular biological statements and observations. For example, the two statements I have used as objects of analysis in transcriptional and chromosomal biology lead to interruptive ambiguity in the interpretative trajectory of these observations in their deductive-nomological form. However, as I have earlier argued this would be consistent with the perpetuation of experiment in the context I delineated. This conceptual framework resembles Kuhn's parameters of what occurs during normal science. Along similar lines, anomaly can be framed as the outcome of a resulting ambiguity that leads to crisis within the puzzle-solving activities of the scientist.

From conventionalism, this methodology borrows the license rationally to accept by convention not only spatio-temporally singular "factual statements" but also spatio-temporally universal theories: indeed, this becomes the most important clue to the continuity of scientific growth (Lakatos 1981: 116). Narrative analysis offers continuity in its spatio-temporal formats in terms of this positive heuristic, "which defines problems, outlines the construction of a belt of auxiliary hypothesis, foresees anomalies and turns them victoriously into examples, all according to a preconceived plan" (Lakatos 1981: 116).

Of course, we can elaborate that science's instinct for metaphors (or maybe scientists' instinct for metaphors), epistemologically grounded in the human condition, (Knudson 2000: 373) may compromise science's claim to lucidity and clarity. Metaphors and images are the basis for continuity in science. Metaphors are inherent in Western metaphysics, which is found to consist of a web of contradictions. However, these contradictions are paradoxically coherent and represent an underlying unification. Thereby, molecular biology, despite its renderings of contradictions and equivocations, may not necessarily undermine itself by resorting to models as metaphors to initiate or represent experiment.

This is not to say that it is the scientific models, or the artistic renderings of scientific data, that are open to critique. The metaphors that these data invokes are derivatives of a culture in which scientists are mediating reality with an unconscious and conscious. Roland Barthes in his seminal essays, "Myth Today," "The Death of an Author" and "From Work to Text," discusses the characteristics of language through an analysis of the field of semiotics, the science of language. He uses myth as an example of how the system of signifier, signified and sign interplays to ascribe a meaning to an object.

He states: "It can be seen in myth there are two semiological systems ... a linguistic system, which I shall call *language-object*, because it is the language which myth gets hold of in order to build its own system; and myth itself, which I shall call metalanguage, because it is a second language, *in which* one speaks about the first."

What Barthes is explicating is that there are two systems of discourse present in any semiological object: one which is explicity present, and another which is implicitly signified and escapes the original intention of meaning of the object in question. Barthes is mainly referring to myths, a type of speech. Barthes states that semiology has taught us that myth has the task of giving an historical intention a natural justification, and making contingency seem eternal.

Scientific language proposes plasticity, not eternal contingencies; it does not prefigure a fixed notion of signifiers. Scientific articles naturalize; however, it is a naturalization open to change. Science avoids mythologizing by leaving things open, and even welcome, to change. Barthes also adds that the "signification of myth is constituted by a sort of constantly moving turnstile which presents alternately the meaning of the signifier and its form, a language-object and a metalanguage, a purely signifying and a purely imagining consciousness." Science, like myth, can be considered to be a metalanguage. To read articles that describe and delineate biological mechanisms, one must use an imagining consciousness. However, it is not a purely imagining consciousness; it is a consciousness that also requires *a priori* factual knowledge of the methodology behind the scientific method. Scientific articles have *three* semiological systems: (1) a linguistic system, which Barthes calls *language-object*, the actual language and terms used and defined by the article; (2) the concepts demonstrated by the article, which is called metalanguage, because as Barthes states, it is a second language, *in which* one speaks about the first; and (3) what I would call the "leading objectivism" of the article, wherein the article promotes the configuration of new approaches to acquiring scientific, factual data.

Chapter 4

Narratology and Scientific Rhetoric: Towards a Semiotics of Scientific Discovery

> An Argument is a Sign, which for its Interpretant, is a sign of law ... a Sign which is understood to represent its Object in its character as a Sign ... What is the essence of Judgment? A judgment is the mental act by which the judger seeks to impress upon himself the truth of a proposition.
>
> —Charles Sanders Peirce, *Logic as Semiotic: The Theory of Signs*

The methods of narrative analysis outlined in Chapter 3 can serve as a springboard for an elaboration of systems of semiosis. In *S/Z* Barthes uses semiotic codes to categorize the narrative elements in Balzac's short story "Sarrasine." His codes are hermeneutic, semantic, proairetic, cultural and symbolic, which, according to Barthes, characterize the plurality of the text. In addition, A. J. Greimas has his own unique analysis of text, which he designates as programs, or segments, which is elaborated upon in a short story by Maupassant, "Two Friends."

The narrative elements as analyzed by semiotic codes are demonstrated in an analysis of a review article entitled "Chromatin Structure and Cardiac Gene Expression" (see bibliography for complete reference). This review article is a major breakthrough in terms of chromosomal dynamics in that it establishes correlations between modification of chromatin structure and gene expression. I have chosen to analyze a review article, rather than a primary scientific publication, because it precisely demonstrates the properties of narrative through its delineation of the organization and sequence of discovery. A review article, such as this one that makes such claims is an ideal choice for semiotic dissection since it elaborates on the construction of a crucial association in a causal and correlative fashion. In addition, how the paper articulates the process of how gene expression is associated with chromatin structure is structured in a narrative fashion. It is easily assimilated to Greimasian semiotics because the argument is made in such a manner that the scientific terms the paper is invested in invokes their binaries. It generates its essential insight by making use of the differential nature of language, i.e. the contrast between dynamic and repressive chromatin. Thus, the paper serves as a most appropriate object to make a pivotal presentation of the Greimsian project. In turn, the connection between chromatin structure and gene expression seemed intuitive to many scientists and investigators attempting to develop the system in which to assert their claims on this subject. In other words, the notion was what Barthes would term an enthymeme, rather than a scientific truth. The writers of this review article conducted research on the cardiac system to provide laboratory evidence for the correlative phenomenon. The review article summarizes the history of discovery in relating gene expression to chromosome structure and seemingly illustrates the process of this phenomena from being enthymeme to acquiring the status of scientific truth. However, their paper summarizing the results of research in this area uses a rhetoric inherent in the scientific arguments which I would like to characterize; their results can be examined in light of its narrative composition vis-à-vis Barthes' and A.-J. Greimas' analyses. In other words, the elaboration of scientific fact is a narrative with its own semiological codes and characteristics that can be demonstrated in a project establishing the analogies between Barthes' analysis of "Sarrasine" and Greimas' analysis of Maupassant; and the review authors recitation of direct proof for the role of chromatin structure in cardiac-specific gene expression.

My main objective in analyzing a scientific publication in semiotic terms is to demonstrate, by dissecting the linguistic properties of a scientific text, that molecular experimentation is highly analogous to the narrative characteristics of literary production. The narrative elements of scientific output can be highlighted, and the significance of ascribing each term in the scientific text to a concept derived from the Greimasian apparatus lies in appreciating that this text is as much a rhetorical system as an elaborate construction of observations and results appealing to objectivity and facticity. As I sequentially explain in this blow-by-blow analysis of "Chromatin Structure and Cardiac Gene Expression," the correlations between semiotics and scientific statements reveal the essence of the elaborate communicative structure behind the methods of laboratory investigation.

Lily Kay ascribes a semiotic structure to the development of the genetic code. She describes how the genetic code obtained a "scriptural iconography" through the use of semiotics, meaning the code was perceived metaphorically as a written text through linguistic imagery. She asks where did this scriptural iconography come from. How did the genetic code come to be textualized? She answers that the discursive and semiotic apparatus of genomic textualization was set in place through a characterization of the historical development of an information-based biology. The code became transformed from syntax to semantics, in Greimasian terms. The dictionary of the genetic code was found to be comma-free and transposable (Kay 2000: 120).

Kay's explication of genomic textualization is a telling example of the use of linguistic tropes in a crucial period of discovery in molecular biology. In a similar vein, I illuminate the relationship between cardiac gene expression and chromatin structure in a review article that highlights the developmental stages in which the discovery of this connection took place. In highlighting this process in terms of the semiotic project, I engage in an archetypal effort. This type of semiotic dissection of a scientific paper (a review article) entails a system in which fundamental syntactical and semantical categories of language, whether on a narrative or discursive level are ascribed to the intrinsic scientific argument of a text. As the intricate terms of the article are set up in a binary, differential level, its conceptual framework is transformed from merely scientific proposition to an elaborate narrative.

The relationship between the sign system as developed by semioticians and its involvement with narrative analysis must be precisely delineated. It was the view of the leading North American semiotician Charles S. Peirce that the "entire universe ... is perfused with signs, if it is not exclusively composed of signs" (Peirce 1991: 448). There is no branch of study or knowledge into which semiotics cannot usefully probe, because there is no branch of study or knowledge which is not sign-based, either because it has evolved a sign-systems of its own, or because it is reliant upon natural language (Sturrock 2003: 79).

In this context, the philosopher Suzanne Langer asserts that language is ultimately a discursive form with expressive value as well as presentational value.

What is the basis for contextualizing the semiotic apparatus in terms of science? When one considers Charles Peirce's definition of signs as "something which stands to somebody for something in some respect or capacity" in the context of the scientific laboratory, one could be correct in stating that experimental practice is imbued with signs. A stain on a protein gel indicates to the researcher an expressed protein; a signal on a developed autoradiogram from a Southern blot points to an expressed gene; a marker on a gel seen on a UV lamp is indicative of a mutant allele. All of these are some of a prodigious number of examples of signifiers in laboratory life. These indicators are visual interdefined units that bear the stamp of providing information that is to be interpreted. Once interpreted, new experiments ensue that develop the narrative of experimentation and provide evidence and data for newly discovered biological facts. Thus, the Peircian notion of an endless sea of interpretation of signs as precursor for biological knowledge can be demonstrated.

The importance of laboratory data serves as the starting point for deductive inference in forming valid biological conclusions and this raw data is in fact related to norms of communication and textual interpretation. Umberto Eco articulates conceptions of signs that fall under the rubric of the role of signs in scientific practice: "A sign is always an element of an expression plane conventionally correlated to one of several elements of a content plane[1]" (Eco 1976: 48). There are three processes involved in sign production: to shape the expression-continuum, to correlate expression with content, and to connect these signs with facts or states of the world. An analogy could be drawn between scientific practice and sign production, of which storytelling and narrative account are a part: The hypotheses which lead to the experimental methods are the elements of expression; as the resulting data are analyzed, interpretations are drawn; and finally conclusions are made based on the interpretations of the data.

Possibly a helpful analogy would be painting or drawing, since we are dealing with non-verbal cues in this context. A good painter has the unusual ability to draw and deals with producing signs that go beyond the ability of the average person, since painting an intricate portrait adeptly of a flower vase is not an equivalent sign to simply pointing a finger at the vase. In a sense, a scientist is like the painter, who has the exceptional ability to perform experiments to produce data that are to be interpreted, as this interpretation would be facilitated by the storytelling function.

The paint and canvas the painter uses are his or her instruments while the scientist has a vast array of technical equipment and techniques at his or her disposal. In this analogy, both produce what are called iconic signs: "An iconic sign is indeed a text, for its verbal equivalent is not a word but a phrase or indeed a story" (Eco 1976: 215). Published scientific articles are like stories or narratives in which an extensive elaboration of iconic signs is incorporated into verbal form in order to demonstrate the veracity of its claims and lead to further discussion, forming a endless chain of semiotic signifiers. The main difference is that for the painter "failure" is usually not a possibility since aesthetic perception is subjective. The scientist is always faced with the possibility of failure when an experiment produces signs which either: (a) do not indicate any-

[1] The expression plane of a communicative system consists of the actual denotoative units in which the system allows the sender to communicate, via the the content plane, or connotative system, the interpretative meaning of the message. A sign function establishes the correlation of an abstract element of the expression system with an abstract element of the content system (Eco, 50).

thing novel, (b) are interpreted as uninformative or irresolvable, or (c) do not lead to further experiments. However, a published paper could be seen as having semiotic properties like a finished portrait or painting. It has pictorial indicators of data, descriptive language of methods and discussion in written form, and informative biological conclusions, all consisting of a visual repository of signs which deserve to be interpreted.

Many semioticians have commented on the nature of physical symptoms in medicine as being demonstrative of signs. When a symptom is associated with a disease, by convention it is semiotically correlated with the diagnosis. Galen was the one of the first clincians to make the correlation between sign and symptom. In a sense certain laboratory data that remain unchallenged or undisputed and are reproducible becomes conventionalized. For example, consider the model of DNA as developed by Watson and Crick of the double helix. The double helix is inextricably correlated with the structure of DNA. As anyone who has seen covers of molecular biology textbooks is aware of, the double helix is, by convention, the DNA molecule. There are numerous examples of conventional data; many are seen in textbooks of molecular science where once data have been conclusively determined, models are drawn depicting the biological processes. The models are visual signs themselves that convey meaning in the form of factual information. Here again the analogy to the painter is relevant. Iconism is in essence a matter of "highly sophisticated semiotic convention" (Eco 1976: 257).

The conceptual apparatus of the sign system that is elaborated above is appropriated by Greimas into a scheme of narratology through his development of the elementary structure of signification: This interpretation of the elementary structure of signification [the semiotic square], while maintaining the principle of binarism, allows for the generation of neutral and complex terms. The elementary structures we have proposed can therefore accept the utterances of scientific, as well as religious and poetic, discourse (Greimas 1987: 174). The elementary structures of signification have immediate relevance to scientific discourse in that

"Persuasive and interpretative are cognitive procedures that, in the first case, end up causing to believe and, in the second, as an act of believing, that is, as an epistemic act and that this epistemic act is a transformation" (Greimas 1987: 167). *Convaincre*, the French term for to convince, leads us to Peirce's citation stated at the outset and means "to lead someone to recognize the truth of a proposition." This transformation allows that the epistemic act, situated at the cognitive dimension of discourse, is indeed a transformation, that is a categorical movement from one "state of belief" to another:

From what is negated to what is accepted,
From what one doubts to what one accepts.

The transformations are situated at the level of deep syntax. The movement from the negated to the accepted can be localized as an operation on the epistemic square that enables the "narrativization" of these transformations. We can apply the general principle of the conversion of structures to them as we move from one level to the other. Thus, a cognitive operation of the logicosemantic type can, in the surface syntax, show a series of narrative programs organized hierarchically. We must not forget that it is at the anthropomorphic level (surface syntax takes place at the anthropormorphic level) that we find the interpretative doing of the subject whom we wish to convince (Greimas 1987: 168).

The property of laboratory data as enveloped within the sign-signifier system engenders a semiotic view of the textual output as a result of that data. Here, explicitly is the association between the analysis of language to narrative properties of a text.

Within the trajectory of my semiotic analysis of a scientific paper, I highlight the narrative programs that lead to this epistemic transformative act of negation to affirmation and assess the significance of the semiotic squares and paradigmatic axis that I construct in terms of how they relate to the overall project of the paper, which I assert to be to develop a general communicative apparatus to convince the reader of the "truth of its propositions, or facts." This apparatus can be precisely dissected through narrative analysis.

In Balzac's "Sarrasine," as analyzed by Barthes in *S/Z*, the main protagonist ignores the clues that La Zambinella, a castrato on the opera stage, is a not a woman by his misguided assumptions of femininity. Barthes terms Sarrasine's misinterpretation of cues as exemplifying the hermeneutic code of Enigma. In turn, the review article attempts to establish the connection between chromatin structure and gene expression as fact derived from specific laboratory data. The method of the article can be likened to the Sarrasinean endeavor: to selectively perceive only the paradigmatic history of facts that establish this correlation as scientific truth. In other words, data can be seen as a hermeneutic code that when perceived solves certain enigmas (however, that perception must be attempted by objective analysis). But as one analyzes the article, its content can be broken down according to the narrative codes of Barthes and a certain rhetoric of scientific methodology can be ascertained in the context of the syntagm of laboratory data and the authors' explanation of research on a paradigmatic scale.

Initially, the abstract can be dissected in terms of hermeneutic, symbolic and cultural codes. It begins: "Gene activation is often proceeded by or accompanied by a perturbation of chromatin structure." This is the HER code, the explicit establishment of the correlation. This is the storytelling code that determines the thematic trajectory of the text. SYM: "Several co-factors of transcription factors have been identified as histone acetyltransferases (components of chromosome structure). [Retinoblastoma protein can form a complex with the acetyltranferases.]" These attributions of the symbolic code are predicated on the antithetical nature of the text. CUL: "However, the role of chromatin structure in cardiac-specific expression has not yet been elucidated. PRO: "The potential significance of chromatin structure in cardiac-specific expression is indicated by [chromatin remodeling factors expressed in the heart] and the demonstration of chromatin remodeling of [developmental and structural genes] during cardiac development." The proairetic code is the action code that indicates the movements of the biological actors in this case. CUL: "We therefore propose to study cardiac-specific chromatin remodeling activity in order to elucidate mechanism controlling [activation] of the fetal heart genetic program." This is the reference to the social network that places the prerogative on discovery.

The prolific nature of these codes in the introductory abstract discloses the shifting points of views and the narrative nature in the establishment of their arguments. The shadow of enthymeme permeates throughout their underlying rhetoric in the proposition of their scientific facts. It transpires from the substantive nature of their assertions to the symbolic code of supporting evidence, the cultural codes responsible for ascertaining their conclusions, the manifestation of proairetic doubt, and finally the imperatives necessary to finding supporting evidence for the reviewers' scientific claims through cultural conditions. The Introduction follows:

> *It was previously thought that the primary role of chromatin was to act as a passive repressor in transcription. However, recent biochemical and genetic evidence indicates that chromatin is a dynamic structure. Upon gene activation, chromatin structure is changed through rearrangement of histones and DNA. Eukaryotic gene expression can be understood only in the context of these changes.*

The article defines itself by a specific representation of gene expression. The oppositions of repression-activation, dynamic-passive and naked DNA-packaged DNA are immediately set up in the introduction. These oppositions form a semiotic square centered around the representation of expression focused on the accessibility of chromosomes (chromatin) to transcription factors.

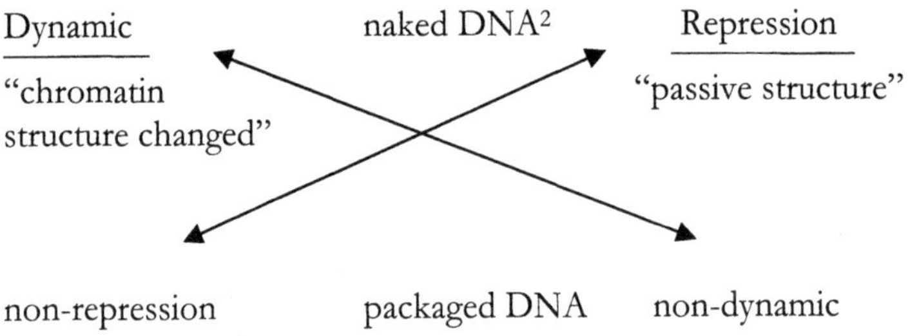

Figure 3 The binaries of a scientific paper as set up in a semiotic square

This semiotic square is based upon the principle that dynamic and repression are binary isotopies that represent differential terms of chromatin structure and gene expression. I.e., chromatin can remain static in packaged DNA and effect repression of gene expression or be dynamic and activate gene transcription.

The article focuses on aligning evidence in favor of the left axiological side of the square. The paradigmatic (a temporal axis) axis of the introduction is formed on the basis of this alignment: *Regulatory elements of several genes were found to be organized with specifically positioned nucleosomes (Simpson, 1986). Such a unique structure has been indicated as an intrinsic part of the regulatory mechanism for eukaryotic genes.* Syntagmatically, the lexemes (narrative units) "dynamic" and "repression" designate the thematic roles corresponding to actantial roles (actantial roles are defined by the position of the actant in the logical linking of the narration and its modal investment, thus making grammatical regulation of narrativity possible) defined by the modalities of chromatin structure: active or passive (Greimas 1988: 110). This is sequence I centered around genetic regulation with its

[2] Naked DNA refers to the state of DNA with respect to the protein components it binds to in a chromosome. DNA is bound to histone proteins in making a nucleosome, or otherwise referred to as packaged DNA. Naked DNA is DNA without the histone proteins.

implications for gene expression. Dynamic is a correlate of a specific chromosomal phenomenom, the rearrangement of histones and DNA. The spatiality and temporality of the scientific representation of chromosomes is established in this introductory paragraph: The spaces, designated as histones, move according to the sequence of the effects of gene transcription on expression, and this movement is necessary for the correlation between gene expression and chromatin structure.

How can we assess the significance of formulating the scientific terms within a framework of the semiotic square? Any proposition formulated by the sender is founded on an epistemic base that moves from refutation to admission. However, this epistemic act, which serves as a prelude to communication, is not just a simple affirmation of self. It is a reaching out, a seeking of consensus, of a contract, which the receiver will either accept or refuse. Between these two instances and these two position takings there is a cognitive space of persuasion and interpretation that, on the level of semio-narrative structures, corresponds to vast structures of manipulation and sanction (Greimas 1988: 171). Thus the semiotic square is the apparatus that mediates manipulation and sanction and characterizes the terms at stake within the binarism of the elementary structure of signification. In doing such, the phenomena elucidated by the paper can be seen as subject to the analytical moves of persuasion that is predicated upon semiotic and narrative constructions, and proof is as much linguistic as scientific.

The contradictory roles of gene activity and repression make implicit narrative trajectories that would establish a passage form one contradictory term to another. The two levels of semiotic representation can be interpreted. The logico-semantic level is the derivatives of scientific data that allow for the conclusory elements of the paper, and the discursive level is the rhetoric inherent in the presentation of scientific evidence. Within the logico-semantic level is the semantic operation that allows for the paper's assertions, wherein the author's conclusions are a function of the demonstrative properties of the text. This is seen in the equation of regulatory elements to specifically positioned nucleosomes, which leads to the conversion of the semantic operations of the text into the discursive level; for it is the activity and positioning of nucleosomes that allow for the correlation between gene activity and chromatin remodeling: *Gene activation is often accompanied by perturbations of the arrays of nucleosomes that make regions of promoters and enhancers more sensitive to nuclease digestion.* The categorization of activation as perturbation marks a discursive transformation on a syntagmatic (co-textual or co-referential) level and forms the isotopy of *nuclease digestion.* Nuclease digestion lies in hypotactic relation to the thematic elements of dynamic chromatin movement through the papers' derivation of *nuclease hypersensitive site.* This corresponds to a complex hypotactic relation since one determiner has multiple determineds. Nuclease hypersensitivity functions as a determiner of gene expression through chromatin remodeling and the spatial representation of the article is constituted by the spatial figure of nucleosomes from which the lexemes dynamic chromatin structure and transcription derive. The semantic explicitation can be illustrated by:

$$\frac{\text{perturbation}}{\text{nucleosomes}} \qquad \frac{\text{digestion}}{\text{nuclease}}$$

Figure 4 Semantic ordering of the scientific terms

This semantic equation, so to speak, explains the correlate between how nucleosomal perturbation takes place through nuclease digestion. It is through the enzymatic activity of nucleases that nucleosomes are able to move. Therefore these axes compose a causal relation.

Nucleosomes are axiologically invested: they are endowed with a certain form of life. "As actors, they can be considered as thematic roles, that is to say, as discursive subjects that can assume narrative trajectories for which the text only indicates possible places of realization" (Greimas 1988: 14). This narrative trajectory follows this path: from the premonition of the correlation of chromatin structure and gene expression to the semantic possibility of chromatin remodeling to its realization in the perturbation of and access to nucleosomes. The inchoative terms are /repression/ and /insensitive promoters/ while the durative terms are /dynamic activity/ and /sensitive nuclease sites/.

In chromatin, nuclease hypersensitive sites are believed to represent "open windows" that allow enhanced access of crucial cis-acting DNA sequences to trans-acting factors. Here, one can undertake a partial paradigmatic exploration of the paper:

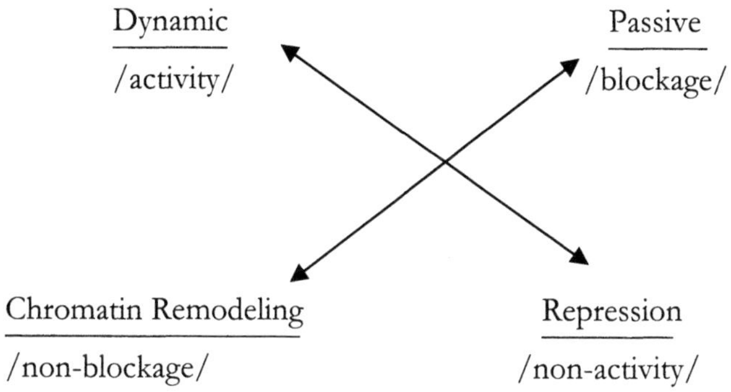

Figure 5 A semiotic square that sets up the isotopies of the paper

This semiotic square is analogous to the first semiotic square constructed. Again the same isotopies are evoked. The paradigmatic terms of dynamism or repression are engaged through either activity or non-activity of the chromatin, or blockage of the sequences from transcription factors. The figurative actors include the underlined items and the items in brackets are invested with axiological values in the sense that narrative correlation is made upon which the paper is predicated. The union of the terms activity and non-activity constitutes the unequivocal existence of the correlation and is hyponymically equated with cis-acting sequences, whose non-blockage is evidence of chromatin remodeling. This leads us to sequence II, which can be connected to sequence I with nucleosomes as the collective actor in SQ1 and cis-acting sequence and trans-acting sequences as the individual actors of SQII. This is a reflexive utterance of actors since it is the remodeling of chromatin through the positioning of nucleosomes that allows for the transcriptive trans-acting factors to act on the cis-acting sequences.

Next the paper acknowledges an equivocation:

> *These accessible regions are defined by their pronounced sensitivity to nuclease cleavage, and are typically two orders of magnitude more sensitive that are other regions in bulk chromatin. A wide variety of functional sequences (e.g. enhancer, silencer, promoter) are associated with hypersensitive sites. Showing that active sites are Dnase-I hypersensitive does not tell us much about the nature of these sites. Indeed, the exact mechanisms for hypersensitivity and its role in gene regulation are only partially understood at present. Based on low-resolution nuclease digestion experiments, it was concluded that these sites represent nucleosome-free regions. However, studies of the MMTV promoter and the albumin gene's enhancer led to a different discovery, i.e., the persistence of nucleosomes with the DHS. Using high-resolution in vivo foot-*

printing analysis, Mcpherson et al. demonstrated that an albumin distal enhancer exists in an array of three precisely positioned nucleosomes only in liver chromatin, where the enhancer is active. Hence, it appears that either possibility (free or persistent of nucleosome) may exist at different DHS in cellular chromatin.

The narrative poles are formed by chromatin: "bulk chromatin" and "cellular chromatin." This program serves to accommodate the acquisition of competence by the narrative subject, since the citation of opposing evidence for differing discoveries informs the prescience of the narrator. The narrative progress is articulated by two distinct categories: the passage from virtual (the subject of wanting) to actual (the subject of doing). The virtualization of the axiology is the establishment of nucleosome hypersensitivity as tantamount to enhancer/promoter sites, and the actualization of the axiology is seen in the ambivalent response to discordant data. The above segment entails a modalization of the collective actor, the nucleosomes, which within the context of actualization of the narrative program includes the

(a) assertion of /free nucleosome-DHS correlation/
(b) assertion of /persistent nucleosome-DHS correlation/

These narrative programs lie at the heart of the essential argument of the paper. If nucleosomes are free, DHS sites are exposed and chromatin remodeling is equated with gene expression. The corollaries to these assertions form the basis for the distinction between preset inducible genes and remodeled inducible genes.

By contrast, remodeling genes are those in which some of the required cis-acting regulatory elements are packaged into nucleosomes (i.e. in a non Dnase hypersensitive configuration). In order for the trans-acting factors to gain access to the cis-acting elements, the nucleosomes must be perturbed in a response to an activation signal; a chromatin remodeling process, which forms Dnase I, hypersensitive sites must occur. Furthermore, it has been demonstrated that the alteration of nucleosomal structure observed in both systems is not due to the consequences of ongoing transcription (Fascher et al., Truss, et al.) Thus, perturbation of nucleosomal structure appears to be prerequisite for, or an initial step in, gene activation.

Sequence 1: introduction to premise of gene expression-chromatin structure correlation
Sequence 2: perturbation of nucleosome inherent in chromatin activity

Sequence 3: inconclusory remarks on nucleosome persistence in DNase I hypersensitivity sites
Sequence 4: perturbation of nucleosome inherent in gene activation

These sequences mark the segments of the narrative unfolding this scientific story.

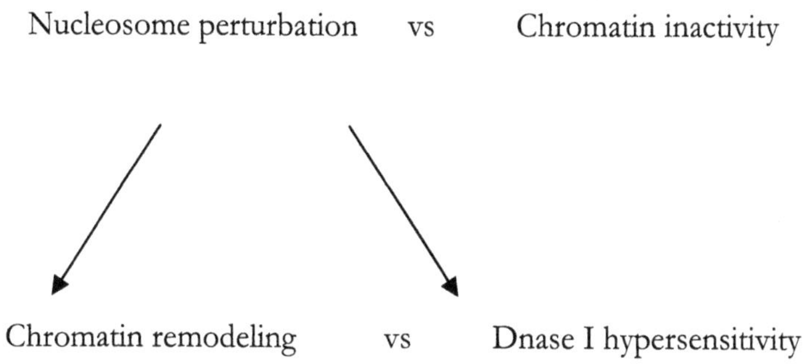

Figure 6 The sequences as part of the narrative programme of the paper

In this context, sequence 3 can be seen as an interpretive halt. The discursive plane evokes how the rhetoric of the paper is established through the conceptual, yet concrete, figurative plane of nuclesomal perturbation, demonstrated through the semio-narrative configuration between syntagmatic and paradigmatic explorations of the paper, which show persuasion and ambivalence.

Discursive plane	chromatin activity	gene activation/ expression
Figurative plane	nucleosomes	perturbation
Narrative plane	Syntagmatic/ Paradigmatic Correlation	Interruptive ambivalence

Figure 7 Segmentation presented as narrative planes

The following three sections are subsets of the syntagm/paradigmatic the themes on the narrative plane under the rubric of "mechanisms of chromatin remodeling":

a)transcription factors bind to nucleosomes
b)modification of histones
c)recruitment of SWI-SNF complex (or other nucleosome remodeling factors)

The process of conversion is apparent here, which enables the passing from one semantic isotopy to another: the category /chromatin activity/ vs /chromatin remodeling/ is a connector of isotopies. The term "chromatin remodeling" is qualified by "transcription factor binding," "histone modification" and "nucleosome factors." Therefore the semic category /dynamic/ vs. /repression/ is a contextual category underlying not only the predicate but the subject of the utterances, from which meaning is formed: *An important event in the remodeling process appears to be initial binding of a positive regulator to its site over a nucleosome; transcriptional factors can also modify nucleosomal structure via recruitment of histone deacetylases; SWI-SNF, chromatin remodeling factors, have been identified.* In addition to the seme /chromatin activity/ the object of cognitive apperception contains one of the category /dynamic/ vs. /repression/:

A = /chromatin activity/ + /dynamic vs. repression/

The semic categories are semantic terms that, again, equate chromatin activity to gene expression within the paradigmatic polarizations of dynamic and repression. However, the category /dynamic/ vs. /repression/ obtains its definitive formulation by the conversion using the metaphoric connector, which is simply the figurative expression of the abstract category /nucleosomal inactivity/ vs. /nucleosomal perturbation/, enabling us to pass form one isotopy to another thanks to the existence of at least one common seme situated at both isotopies at the same time (Greimas 1988: 89). The term "chromatin remodeling" could be substituted for the seme /nucleosomal perturbation/ by the fact that "chromatin remodeling" contains the seme /nucleosomal perturbation/ and the seme /dynamic/. This semantic representation brings into existence a certain cognitive reading since the structural trajectory makes explicit analogies between the analogous semes. In Sequence 7, "Chromatin remodeling of the cardiac myosin heavy chain genes," the isotopy /Dnase hypersensitivity site/ is displaced onto the isotopy /transcription/.

The pragmatic program begins:

> *Several cardiac genes have been analyzed for Dnase I hypersensitive sites. They are chicken light chain gene (MLC2A), cardiac actin gene and cardiac alpha and beta myosin heavy chain genes(MyHC). Using nuclei isolated from chicken embryonic heart, Winter and Arnold (1987) first identified four cardiac-specific DHSs in the MLC2A gene and concluded that these sites may be related to the expression of the gene.*

Here we can say that what characterized the first state, the correlation between chromatin activity and gene expression, made possible the subject of nucleosome perturbation, as the key correlate, whereas the second state is characterized by the actants responsible for nucleosome perturbation conditioned by the cognitive state i.e., /knowing/ made possible through scientific evidence. The modality of /chromatin activity/ actualized becomes a state of realization at the moment of the act of relating, and appears here as a transforming doing, situated on a cognitive dimension: DNase I hypersensitive sites are responsible for gene expression in cardiac genes.

However, there is a disjunctive hesitation that follows, which serves as a primer for the assertion of the pertinent scientific facts and data providing evidence for them:

> *Although these studies demonstrated the existence of DHSs in the cardiac-specific genes, they did not address whether, during mammalian cardiogenesis, there is a change in chromatin structure accompanying cardiac-specific gene expression.*

This hesitation foreshadows the ultimate conclusions of the text, as we shall see later.

We see here two trajectories, the general and the specific:

General trajectory = "chromatin remodeling/nucleosomal perturbtion"

Specific trajectory "DNase I hypersensitive site/cardiac gene expression"

Figure 8 General and Specific Trajectories

By assigning these terms to the general and specific trajectories respectively, chromatin remodeling is a general manifestation of DNase hypersensitive sites which specifically effect nucleosomal perturbation. In essence, without the association of DNase hypersensitive sites with nucleosomal perturbation, cardiac gene expression through chromatin remodeling would not take place. A structure of communication underlies the both trajectories, however the "specific trajectory" is composed of the cognitive program which is superimposed on the pragmatic program: *In order to explore the role of chromatin remodeling in cardiac gene expression, we studied DHS pattern of the hamster cardiac MyHC genes during cardiac development.* It transforms the /non-making known/ to the /making known/, which constitutes the persuasive doing so that the plane of interpretive doing lies within existence rather than non-existence. However, this interpretive doing can be considered in two different ways: either as the terminative aspect of durative temporality or as the inchoative aspect of a new temporalized doing (Greimas 1988: 100). In other words, DNase I hypersensitive sites serve as a primer for the elaboration of nucleosome perturbation in transcription events in early cardiac development, the system which now is the determinant subject. The cognitive program consists of the events in cardiac development vis-à-vis transcription activity. In short, the regulatory elements of cardiac alpha and beta MyHC genes were identified as playing a role in the regulation of cardiac MyHC gene expression:

> *Consistent with transgenic studies, we identified a cardiac-specific DHS located in the 4-kb upstream region of the cardiac alpha-MyHC gene... Within this site we found a conserved GATA-motif that interacts specifically with GATA-binding factors in nuclear extracts of cardiomyocytes at various developmental stages. These data provide further evidence to support the role of GATA factors in the regulation of cardiac alpha-MyHC gene expression.*

The individual universe of /GATA-factors/ vs. /cardiac alpha MyHC expression/, constituting the abstract representation of regulation of gene expression in cardiomyocytes, is homologated with the collective universe of /chromatin remodeling/ vs. /transcription activity:

> *Two cardiac specific DHSs were identified within the 5-kb upstream region of the cardiac beta-MyHC gene. Although these two sites were readily detectable using nuclei from neonatal hamster heart, the proximal promoter site disappeared when adult hamster heart nuclei were used. This alteration in the DHS pattern corresponds to a shift in expression of the cardiac MyHC genes in the hamster. This suggests [that the existence] of this DHS in the heart could be the result of tissue-specific chromatin remodeling activity.*

The figurative axiological structure would be as follows (formed from the reduction of discursive isotopies as a thematic role):

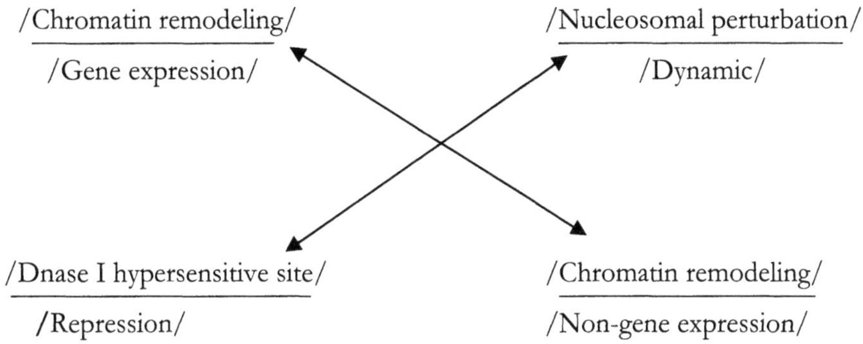

Figure 9 On the discursive plane, the terms of the paper set as figurative axiological structures

Again, this reiterates the pattern of semiotic square construction in the establishment of the isotopies that equate DNase I sites with nucleosomal perturbation and chromatin remodeling with gene expression. The next sequence, "Cardiac transcription factors and chromatin remodeling," distinguishes two types of knowledge in the establishment of narrative competence: narrative knowledge and thematic knowledge:

> *The cardiac transcription factors play a pivotal role in cardiogenesis and cardiac-specific gene regulation. By using gene targeting analysis, cardiac transcription factor GATA-4 has been shown to be essential for cardiac development. Several lines of evidence indicate that GATA-4 ... mediates the reactivation of beta-MyHC gene in a cardiac hypertrophy model. More significantly, GATA-4 was recently shown to link to NF-AT3, a downstream transcription factor of calcineurin signaling pathway...In order to activate fetal gene expression, cardiac transcription factors first have to overcome the chromosomal barrier ... Thus it is possible that synergistic activation of the ANF (atrial natriuretic factor) promoter by GATA-4 and NF-AT3 results from a similar mechanism in which NF-AT3 is recruited to facilitate GATA-4 to bind to its site over nucleosomal DNA. However, this model remains to be tested.*

The syntactic articulations takes place under the sign of iterativity, or isotopies which allow for the coherence of the story:

(1) pivotal role in cardiogenesis and *cardiac specific gene regulation*
(2) GATA4-reactivates beta-MyHC gene
(3) Cardiac hypertrophy
(4) Activate fetal gene expression, cardiac transcription factor overcomes chromosomal barrier
(5) Mechanism facilitates GATA-4 to bind over *nucleosomal site*

This is examination of the iterative property of isotopies that acknowledge the referential network of the text. From a textual point of view, there is a pivot between cardiac-specific gene expression and nucleosomal sites. Therefore two types of activity oppose each other, marked by transcription in cardiac fetal development and cardiac hypertrophy. The last articulation marks the development of a positive deixis and a negative deixis:

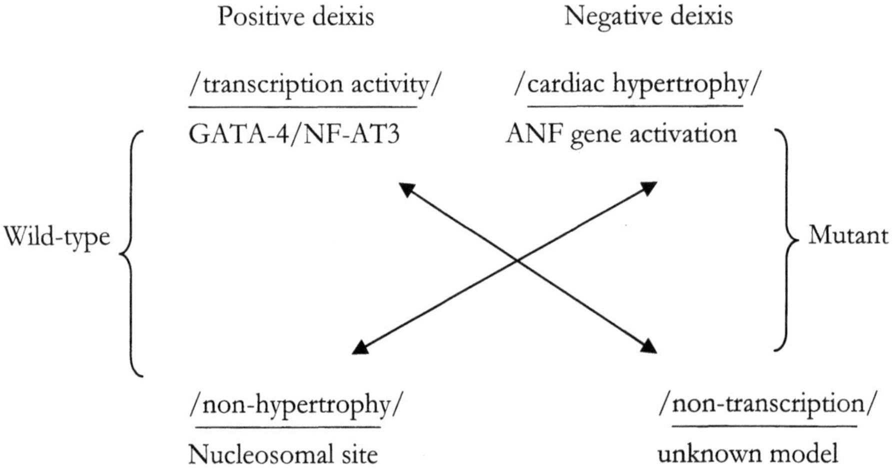

Figure 10 Positive (favorable) and Negative (unfavorable) outcome

At this point, the text introduces new isotopic parameters in which the novel argument is constructed that transcription activity is predicated on changes in cardiac nucleosomes during development of the adult heart. In the mutant heart, these changes do not take place. The construction of the Narrative Program requires knowledge and a "knowing how to do." Narrative competence, the "knowing how to do" about the construction of new narrative objects, presupposes a knowledge about the modes of existence of the Narrative Program (Greimas 1988: 176). Narrative competence lies in the "syntagmatic intelligibility" of the text, in this case the correlation between nucleosomal perturbation and transcription and chromatin remodeling and cardiac gene expression. Narrative knowledge comprises the specific trajectory of cardiac fetal gene transcription and cardiac hypertrophy, which is interwoven within the thematic knowledge of gene expression and chromatin structure, almost tantamount to the general trajectory of the text. The narrative knowing-how to do projects new virtual programs at each step: the identified goal of gene expression and chromatin structure is the source of the program to which the narrative utterances return. Transcription and nucleosomal perturbation are situated as objects of value on the "axis of wanting." Through continuous syntagmatic linking, the gap separating the object of value from the subject, or the identified goal, is filled in by sub-programs involving DNase I hypersensitive sites, chromatin remodeling, cardiac fetal gene transcription and cardiac hypertrophy: The coherent NP can be constituted from these types of knowledges, and the interpretative doing can be derived from an analysis of the basic narrative program:

> *Whether chromatin remodeling is applied to other fetal cardiac genes remains to be determined. It might be interesting to investigate the chromatin structure of the ANF gene, since the developmental expression pattern of this gene is somewhat similar to that of cardiac beta-MyHC gene … Since activation of fetal cardiac genes is the target of hypertrophic signals, cardiac-specific chromatin remodeling factors should act as essential factors in response to hypertrophic signals. Thus identification of the cardiac-specific chromatin remodeling factors and determination of which cardiac nuclear factors are capable of recruiting this chromatin remodeling machinery will be crucial to our understanding of the reactivation of fetal genetic program in cardiac hypertrophy.*

The establishment of the discourse is finalized in this last paragraph. The discourse within the discourse, i.e. gene expression and chromatin remodeling and the factors responsible for both, takes place on a rhetorical plane within the narrative program. The rhetorical plane shifts the program from chromatin remodeling activity to fetal genes to cardiac hypertrophy: *This cardiac-specific chromatin remodeling activity may be an alternative therapeutic target for preventing (or slowing down) the progression of cardiac hypertrophy.* The final utterance by the subject, enunciated through probability and likelihood, opens up the door for new avenues of investigation.

Chapter 5
Historical and Sociological Frameworks from Narrative and Semiotic Perspectives

One approach that might benefit history of science is the semiotic one. Historians of science have commented on the discursive and rhetorical properties of scientific communication and attributed narrative elements to scientific practice (Golinski 1998: 103, Rouse 1990: 179). Semiotics offers an ahistorical set of analytical tools which we can apply to specific historical contexts, not indiscriminately, but in an accurate and precise manner that takes into consideration historical perspectives. Drawing on Bastide's work, my aim is to look at debates in a semiotic way. In this light, this inquiry can be broadened to address the views of sociologists and historians of science. Steven Shapin and Simon Schaffer offer an exploration of how experimental matters of fact are constructed in their work *Leviathan and the Air Pump*:

> If one wanted to produce authenticated experimental knowledge—matters of fact—one had to come grip with this space and work in it with others...The laboratory, therefore, a disciplined space, where experimental discursive and social practices were collectively controlled by competent members. In these respects, the experimental laboratory was a better space in which to generate authentic knowledge than space outside in which simple observations of nature could be made (Shapin and Schaffer 1986: 39).

Leviathan and the Air Pump concerns a historical assessment of the debate that concerned how matters of experimental fact were legitimately reached, especially in the context of natural philosophy during seventeenth-century England. Shapin and Schaffer argue that this debate was as much a game of language as it was an argument over the epistemological concerns of what constituted the fundamental source of knowledge.

Shapin and Schaffer argue that the experimental matter of fact rests upon causal explanation. In this sense, I would argue that one clear dimension of the argument between the natural philosophers and the experimentalists is one of language and use of terms. In this context, narratology can address an historical representation of debate, i.e., between Boyle and Hobbes on arriving at matters of fact and natural philosophy because one of the key parameters of the debate was causation. Boyle needed to distinguish between his experimental apparatus as causes and about their causes: "There was to be a crucial boundary between the experimental matter of fact and its ultimate physical cause and explanation" (Shapin and Schaffer 1986: 51). Shapin and Schaffer argue that Boyle's causal talk was "grounded in convention." From the perspective of narratology, I argue that causation, termed *fabula* is a fundamental characteristic of how events are chronologically determined. While Boyle based his experimental philosophy upon matter of facts as being a social as well as an intellectual category, and the validation of his experimental performances was based upon the virtual witnessing through literary technology, through narrative analysis we can see Boyle's experimental philosophy as an elaboration of events that causatively account for each other.

Shapin and Schaffer argue that the fundamental distinctions that Hobbes discerned in his criticism of experimental philosophy were made in the political and historical context in which Hobbes perceived of the "danger" in the language formulation of the vacuum. However, I would argue that the debate is as much rhetorical, and the terminology of the conflict can be formulated semiotically. In semiotic terms, the conflict would look very different. It would be about the binaries of vacuum and plenism, terms that emerge from the differential nature of Boyle's and Hobbes' linguistic appropriations. To understand debate is to recognize the binary distinction between vacuum and plenism, which are as much linguistic signifiers as they are experimental or philosophical categories. Boyle demonstrated the existence of a vacuum through his air pump, and the right philosophy to direct its existence was predicated upon the proper use of language. Hobbes, on the other hand, argued that the issue of pneumatics was an issue of *right language* in philosophy (italics Shapin and Schaffer, Shapin and Shaffer 1986: 87). As Shapin and Schaffer imply, the work of the laboratory translates into its manifestation in language. "Consider also the language Boyle used to describe to his principal ontological concern: the air and its properties of spring, weight, and pressure" (Shapin and Schaffer, 1986: 52).

In this context, the narrative programme highlighted by Bastide illuminates the parameters this debate. The transition from *faire-savoir* to *faire-croire* lies at the heart of how Hobbes appropriates language to evaluate the credibility of the vacuum, and thereby the air pump. If experimental matters of fact can lead to accurate statements about reality according to Boyle, (*faire-savoir*), Hobbes would counter that this constitutes an improper, and dangerous, use of language, particularly the language of natural philosophy, and does not accede to belief (*faire-croire*).

These semiotic and narratological approaches can help us understand another case study in the history of science. Naomi Oreskes articulates a moment of discovery in plate tectonics, when the theory of continental drift belatedly emerged as accounting for movement of the earth's continents. Oreskes describes the initial negative reaction in America to Wegener's theory of continental drift and accounts for this reception on historical grounds, distinguishing between American and European geologists. Oreskes states that the American reaction was in part due to "resistance to causal theorizing" among American geologists, and aversion to Wegener's method of explanatory inference (Oreskes 1999: 134-135).

It could be argued that Wegener did not have the rhetorical devices to convince his audience into accepting his theory. Wegener's theory was rejected because, as Oreskes documents, it counteracted prevailing scientific approaches at the time. It was deductively constructed and, in its universality, it countered American ideals of pluralism and democracy. Narrative analysis may provide some additional insight into the complexities of why the theory of continental drift was rejected. The cause and effect manner by which Wegener approached his theory would have been read as fallacious by his educated readers if indeed they interpreted it as a correlation rather than a valid cause-and-effect relationship.

Narrative highlights this process:

> Narrative by its arrangement of events gratifies our need for order, of which perhaps the commonest kind of order...is the perception of cause. If this can make narrative a gratifying experience, it can also make it a treacherous one, since it implicitly draws on an ancient fallacy that things that follow other things are caused by those things...A correlation does not establish a cause and effect relationship. However, cause and effect work sequentially, just as stories do. A cause can in fact be any number of things, or any combination of things, that precede an effect, not necessarily the thing that the narrative draws our attention to (Porter-Abbott 2002: 38-40).

Historical evidence, as Oreskes points out, indicates Wegener acted upon inspiration in which he deductively arrived at his theory, rather than through facts produced by hard work and observation. As a result, his American colleagues initiatlly rejected the theory. Further, the Wegener's narrative framework of implied causative agency among the continents, which seem to fit together, may have been construed as correlation, making it a "treacherous" narrative. Wegener may have been guilty of, in Porter-Abbott's terms, writing a narrative so uncluttered by competing elements, that the genuinely demonstrable cause-

and-effect elements of his geological story was misconstrued as correlation. Americans were widely committed to the method of multiple working hypotheses—the idea that scientific evidence should be weighed in light of competing (multiple) theoretical explanations, which one held provisionally until the weight of evidence was sufficient to compel assent (Oreskes 2003: 10).

This analysis must be qualified because Boyle's historical context was much different from that of Wegener. However, as long as we stay within the specific context, we can use semiotic analysis to provide other perspectives in evaluating crucial debates in history of science.

Furthermore, studies on the social constructivism of science have appropriated semiotics in their articulation. It has been argued that Latour and Woolgar's ethnographic study of laboratory science has been formulated in semiotic terms. Latour and Woolgar articulated the "actor-network" theory that posited how material objects determined the persuasive properties of a scientific text. Jan Golinski elaborates on how Latour and Woolgar's use the term "actant," that is derived from semiotics. "The first chapter of *Science in Action* is devoted to the rhetoric of scientific writings. There, Latour decodes the dense undergrowth of references that usually accompanies a scientific paper as a means of buttressing the author's statements with the authority of predecessors and colleagues. Scientific rhetoric is a stylized way of enhancing the credibility of written assertions by raising the cost of dissent. To resist the claims of a well-documented paper is to pick an argument with all the authorities it cites. In Latour's view, facts and machines are transmitted in the same manner, passed down the chains of linked persons and things that he calls "networks." Networks are fundamental to the motion from local knowledge in the laboratory to worldwide 'technoscience'...By extension of these networks, the world is seen to be reshaped to resemble the special settings in which experimental artifacts are made...For Latour, an understanding of technoscientific research requires that we understand the agency of nonhuman as well as human beings. Latour's networks are heterogeneous linkages of people and things—associations of human and nonhuman entities, to which he gives the common name 'actants,' borrowed from semiotics" (Golinksi 1998: 39).

Mapping the construction of these heterogenous networks of human and nonhuman actants offers a way of following the passage form controversial to accepted knowledge (Golinski 1998: 40). This construction is analogous to the semiotic analysis, whereby the transition from doubt to acceptance follows the pattern of narrative thematization on discursive and figurative planes.

Golinski continues that "At times, Callon and Latour seem to be advancing their actor-network theory as a kind of semiotics, in other words, as an account of how the actants they identify function as signifiers in a discursive field. In the discourse of scientists and engineers, effects may be assigned to humans or nonhumans, and the two classes of entities may exchange signifying roles" (Golinski 1998: 40)

As a counterview to Callon and Latour's assessments, social constructivists of science have emphasized socialization as the driving force behind science. David Bloor and Barry Barnes showed how Thomas Kuhn's work could yield valuable resources for building a constructivist sociology of scientific knowledge. [Barnes and Bloor] have argued that the notion of a paradigm as a concrete exemplar—a model problem solution—suggests a pragmatic alternative to the traditional philosophical view that science is governed by a logical structure of theory...If paradigms are seen primarily as models, then science appears as an enterprise of practical reasoning governed by accepted conventions rather than by logical deduction form some theoretical structure (Golinski 1998: 13-14). The view of the way science proceeds in line with the position that Bloor and Barnes describe as "finitism." Bloor defined finitism as "the thesis that the established meaning of a word does not determine its future applications...Meaning is created by acts of use (Bloor 1983: 25). Barnes added, "Finitism denies that inherent properties or meanings attach to concepts and determine their future correct applications; and consequently it denies that truth and falsity are inherent properties of statements" (Barnes 1982: 30-31).

The appeal of this view for Barnes and Bloor was that it made science look more like other aspects of culture. As a kind of craft, science appears more open to the naturalistic approach to understanding its construction; the epistemological barriers around it are lowered. It is then plausible to regard scientific beliefs as being inculcated by relations of authority within educational institutions and sustained by conventions in specific communities (Golinski 1998: 17). Barnes wrote, "paradigms, the core of culture science, are transmitted and sustained just as is culture generally: scientists accept them and become committed to them as a result of training and socialization, and the commitment is maintained by a developed system of social control" (Barnes 1985b: 89).

According to Bloor and Barnes, social relations determine the outcome of a scientific endeavor, as opposed to the hypothetico-deductive method and its rhetorical appropriation. However, I would argue that semiotic analysis can qualify the notion of finitism. I have articulated that signs are an ever-pervasive entity in the molecular biology lab as a result of techniques employed there, and that the transformation of the procedural method of scientific experimentation into accepted discourse is a function of the actantial properties of the text. According to this examination of scientific communication, a scientific term evokes a binary; by negating its antithesis, the rhetorical force of the term becomes manifest. I would argue that semiotics may not totally discount social agency and relations in the transmission of the meaning of scientific statements. Indeed, semiotics renders forms of authority to which Barnes refers into a persuasive modality. However, while metaphysical speculations as to the veracity of the claims presented by the text may not be resolved by such, semiotic analysis attributes potency to the differential nature of language and asserts that the meaning of a term inherently precedes and supersedes its cultural appropriation. Furthermore, semiotics does not contradict the universality of scientific statements and supports the argument that science cannot be reduced to social phenomena. Semiotics accounts for scientific phenomena, as I have argued, as a result of structures of language that preexist and overdetermine social and cultural forces, a premise of structuralism. The premise of my analysis is that the hypothetico-deductive model undergrids the practice of science. This would predicate that science concerns logical deduction from theoretical formulation.

However, in the context of semiotics, science, if not purely a social construct, may to some extent be considered a sociolinguistic construct. Callon and Latour argue that purely social relations seem too weak to do the job: To claim that what determines the outcome of a controversy is a decision to accept one set of evidence rather than another—a decision based upon the social ties among individuals in the community—is to rest a purported explanation upon ground far too shaky to support it (Callon and Latour 1992: 352-356).

Golinski points out as well that it is precisely the power of narrative that allows for both sides of the debate to posit their arguments: "As sociological debates about nonhuman agency have raised issues for historical narrative, so narrative has been seen as an arena in which solutions to the theoretical problems may be worked out. It is also striking that two sides of the debate between Callon and Latour and Collins both pay the tribute to the historians Shapin and Schaffer of enlisting their work in support of their claims. Perhaps this is because the historians' rich narrative succeeds in conveying the experience of the openness of scientific practice and of its engagement with material objects" (Golinksi 1998: 45).

In other recent work, Joseph Rouse has attempted to apply to science itself the category of narrative, claiming not simply that it characterizes a style of writing but that it constitutes "the temporal organization of the understanding of practical activity." In their day-to-day work to impute meaning to their own and others' actions, practicing scientists are constantly engaged in storytelling, interpreting the direction of development of past work and projecting it into the future." Scientific research is social practice, whereby researchers structure the narrative context in which past work is interpreted and significant possibilities for further work are projected (Rouse 1990: 179). Events are only given meaning by their incorporation into narratives, which individuals are constantly competing to complete. The narrativity of practice is a consequence of the embeddedness of all meaningful action in sequences of events understood as stories. "To act," Rouse proposes, "is to be ahead of oneself, that is, to have some understanding of what would be to have done the action in question." All actions "are being enacted toward the fulfillment of a projected retrospection, but one which is constantly open to revision, as befits a story not yet completed" (Rouse 1990: 183-184).

According to Thomas Kuhn the process of scientific change is actually based on non-objective, socio-political factors (Aronson 1984: 104). Scientific progress must be viewed not in logical but human terms, in particular, terms of the replacement of a total network of values, commitments and theoretical institutions of a scientific community with an entirely different set (Aronson 1984: 104). As mentioned previously, Kuhn articulated the concept of the scientific paradigm. Among the scientific activities of which paradigms as networks of rules and commitments are to make sense: designing experiments, making calculations and predictions, interpretation of data, and "puzzle-solving" or assimilating seemingly disparate phenomena into the theory (Aronson 1984: 109).

However, it has been argued that science is driven by theory. On the scientific level, the tentative adoption of a new conjecture or theory may solve one or two problems, but it invariably opens up many new problems; for a new revolutionary theory functions exactly like a new and powerful sense organ. If the progress is significant then the new problems will differ from the old problems: the new problems will be on a radically different level of depth. This happened, for example, with Einstein's theory of relativity; it happened with quantum mechanics; and it happened with molecular biology. In each of these cases, new horizons of unexpected problems were opened up by the new theory (Popper 1981: 83).

In contrast, the great scientific achievements are research programmes which can be evaluated in terms of progressive and degenerating problemshifts; and scientific revolutions consist of one research programme superseding (overtaking in progress) another. This methodology offers a new rational reconstruction in science in that genuine progress in science is cumulative and takes place on the ground level of "proven" facts; the changes on the theoretical level are merely instrumental (Lakatos 1981: 111).

Other historical viewpoints have been articulated by Ian Hacking. According to Hacking, we cannot tell whether a research programme is progressive until after the fact (Hacking 1981: 118). Hilary Putnam hopes, like Peirce, that there is a scientific method upon which we shall come to agree, and which in turn will lead us all to agreement, to rational, warranted belief. However, Lakatos argues that there is no forward-looking rationality, but we can comprehend the objectivity of our present beliefs by reconstructing the way we got here (Hacking 1981: 119).

Postmodern visions of science have also been articulated, as exemplified by Paul Feyerabend, who has a more radical view of science. Knowledge so conceived is not a series of self-consistent theories that converges toward an ideal view; it is not a gradual approach to truth. It is rather an ever-increasing ocean of mutually incompatible alternatives, each single theory, each fairy-tale, each myth that is part of the collection forcing the others into greater articulation, and all of them contributing, via this process of competition, to the development of our consciousness (Feyerabend 1975: 21).

Taking extreme sides, these viewpoints seem to be incommensurable. If science is reduced to myth or fairytales, how can progress be accorded to its practice? Furthermore, if scientific progress functions by hindsight, how can a method be attributed to it? And is this method constructed by consensus? Is science cumulative, or driven by theory?

By positing these questions, we can ascertain how narrative and semiotic analysis allow for a unique conceptualization of theoretical formulations that center on these debates about how science functions. In this evaluation, semiotics does not inherently refute the cumulative nature of science, nor negate the notion of a theory-driven, progressive one. Semiotics offers a precise method for demonstrating how scientific texts can be brought to bear textually and delineates the rhetorical pattern of scientific communication. To be put on one side or the other in this debate may be beyond its purview. However, semiotics does give precedence to a formulation that does service to its structuralist premises. While semiotics may not editorialize on the progressive versus cumulative nature of science debate, I would argue that a theory-driven science that is determined by logical predicates over socio-cultural authority is most in accordance with semiotic analysis. I do so since, in semiotic terms, a scientific theory can be considered a linguistic construct that can be attributed a differential character. While my analysis must be delimited by the scope of the semiotic method, the main objective here is that semiotic and narrative analysis can illuminate certain perspectives in the studies concerning the history and sociology of science.

Conclusion

The specific trajectories of this text are:

(1) That the elements of narrative (time, causation, plurality, segmentation, ambiguity and open-endedness, and binary structure) can be ascribed to the routines and procedures of the laboratory, particularly in transcriptional biology and chromatin structure;

(2) That the perspectives in which one conceives of statements in transcriptional and chromosomal biology can be illuminated and addressed by the narrative methods of Barthes and Greimas;

(3) That scientific communication can be dissected semiotically and its persuasive capacities can be highlighted as a rhetorical operation.

However, the "big picture" questions remain in terms of how reliable, robust scientific knowledge is generated. The image of the monolithic scientific method has been challenged by historians of science, who see different methods and approaches in arriving at scientific knowledge, and acknowledge that there are cultural, artistic, and political questions that inspire and motivate scientific work. The view of science as a progressive enterprise that moves closer to truth is problematic, according to this perspective. There exists a confident response to this view, termed present-day exceptionalism that refers to the present day as "perfect" and that the past as erroneous. However, there is also the view of pessimistic metainjunction that states: if previous facts are erroneous, how do they reflect upon present-day facts?

In this respect Ian Hacking highlights:

> One can see that by direct inspection that knowledge has grown. This is not a lesson taught by general philosophy or history but by detailed reading of specific sequences of texts. [There] is no arguing that some historical events do exhibit growth of knowledge. What is needed is an analysis that will say in what this growth consists, and tell us what is the growth that we call science and what is not (Hacking, 119).

Hacking offers a textual paradigm for assessing growth of knowledge in science, a paradigm for which narratology can provide insight and may possibly foreground the analysis that Hacking proscribes. By highlighting the narrative properties of a scientific statement as a textual sequence, one can determine the statement's access to knowledge through its connotative plurality.

One of the central terms in this study is the "scientific method." I discussed Carl Hempel's formulation of the deductive-nomological method for narratively evaluating statements in transcriptional regulation. However, philosophers, sociologists, and historians of science have debated over what constitutes the scientific method, and whether or not there are a plurality of scientific methods. Kuhn has argued for scientific revolutions that account for wholesale changes in scientific activity, while Latour has stated that laboratory fact is a social construction in which scientific data acquires credibility through negotiation among a body politic of scientists. The extremes of this debate may be represented by the relative counterpositions between Karl Popper and Paul Feyerabend.

Popper was one of the crucial formulators of attributing a specific methodology to the scientific endeavor, and believed in the unity of a scientific method. He argued in the introduction to his *The Logic of Scientific Discovery* that a scientist puts forward a series of statements and empirically constructs hypotheses and tests them against experience, observation and experiment (Popper 1959: 7). In contrast, Paul Feyerabend in his *Against Method* states that some of the most important formal properties of a theory are found by contrast, not by analysis. A scientist who wishes to maximize the empirical content of the views he holds and who wants to understand them as clearly as he possibly can must therefore introduce other views; that is, he must adopt a *pluralistic methodology* (italics Feyerabend, Feyerabend 1975: 21).

Narrative theory can find its correspondences in both sides of the debate. Certain philosophical discourses of science correspond to the principles of narrative and particular genres of narrative. As I have argued, the perpetuation of molecular biological statements occurs through the principles of narrative, which in turn are accounted for by the causality of these statements. For example, Karl Popper has argued that "the principle of causality is the assertion that any event whatsoever can be causally explained—that it can be deductively predicted." He notes that "the game of science, is, in principle, without end. He who decides one day that scientific statements do not call for any further test, and that they can be regarded as fully verified, retires from the game" (Popper 1959: 61, 53). In this context, narrative analysis depicts scientific communication as open-ended, and, when applied to the scientific endeavor, incurs its perpetuation. Causation is a central property of narrative and resembles Popper's principle of causality. The narratological assessment of molecular biological statements is consistent with this characterization.

However, according to Paul Feyerabend, science is an essentially anarchic enterprise, and science sustains itself through this anarchy. The constitution of Feyerabend's statements has a literary correspondence in the modern narratives of William Faulkner, (*The Sound and the Fury*, *Absalom!*) Samuel Beckett (*Malone Dies, Molloy*), and James Joyce (*Ulysses*), whose modes of conveying plot and theme challenged traditional notions of how narratives progress. Faulkner, Beckett, and Joyce were instrumental in advancing what was to be called the "modern novel," whose organization lacked the logic of temporality, was "disorderly", and could be described as "anarchic." In this context of Feyerabend's account, the task of the scientist and the unconventional novelist begin to seem similar. In turn, Barthes' formulation of the plurality of meanings in a text is consistent with this view as well.

The implications of the preceding analysis are that contrasting positions on the nature of the scientific method have their rough approximations in theories of narrative and in textual genres. This points to the overarching and universal nature of the characteristics of narrative, a point noted in the introduction (cf. Barthes' citation). In its totality, narrative analysis offers an interpretive framework to assess how literary texts find their correspondences in certain discourses in the philosophy of science. The properties of narrative, causation, temporality, and ambiguity can be attributed to articulations of how scientists conduct laboratory work.

Narrative analysis aims at evoking elements of realism in a text, and Ian Hacking discusses scientific realism and anti-realism in the representation of science: "Science is said to have two aims: theory and experiment. Theories try to say how the world is. Experiment and subsequent technology change the world. We represent and we intervene. We represent in order to intervene, and we intervene in light of representations. [Debates] about scientific realism are couched in terms of theory, representation and truth" (Hacking 1983: 31).

Scientific realism can be considered similar to textual realism. In the context of reading, "realism" appears to be a broad area of narrative without any identifiable conventions, one in which literary artifice has disappeared and everything happens as it would in life. In the best realistic narratives, we are startled into awareness of the real: we would never have imagined the revelation that came just after we turned the page, but after it appears, we realize that it was inevitable—it captures a truth of experience that we dimly realized, however dimly all along (Martin 1986: 58). Narrative theory finds some regularity in the attitudes on which belief is based, including credulity, credence and skepticism. Scientific realism says that the entities, states and processes described by correct theories really do exist. The realist holds that we often get close to the truth. Definitions of "scientific realism" merely point the way. It is more an attitude than a clearly stated doctrine.

The essence of the dichotomy of scientific writing versus literary writing lies in the long-held and vigorously supported argument that both a literal language and figurative language exist side-by-side (Neussel, 487). In this traditional typology of language into two different categories lies the essence of the claim for distinct scientific and literary languages. I have tried to establish in this study that there are parallels between the literary and scientific texts, and thus narratological analysis can apply to both. I have also tried to show how literary theory, particularly French narrative theory, can be applied to scientific communication.

The metatheory of narrative analysis that I have outlined serves as a foundation for an apparatus for evaluating the biological moves in the laboratory. The results of scientific investigation, the scientific publication, possess narrative units that function to accord levels of meaning within scientific statements. I show how the scientific narrative must be constructed on a connotative level. Narrative analysis gets at the heart of scientific controversy because it offers a reflection of the biological moves in the lab and a metatheoretical apparatus for assessing conflicting trajectories of scientific investigation. Using Barthes as a starting point, one can infer that scientific language proposes plasticity not eternal contingencies; it does not prefigure a fixed notion of signifiers. Scientific investigation is perpetuated by an effort that evokes narrativity. Scientists provides the impetus for furthering their research through the process of narratology. The scientific text produces ambiguity and open-endedness; for scientific conclusions are tentative and the effort of a laboratory is predicated upon data emerging or changing as a result of another lab's work. However, further, scientists think of original perspectives to think of biological systems through a process that can be illuminated through narratology.

The narrative elements as analyzed by semiotic codes are demonstrated in a scientific review article. The objective in analyzing a scientific publication in semiotic terms is to demonstrate, by dissecting the linguistic properties of a scientific text, that molecular experimentation is highly analogous to the narrative characteristics of literary production. The sequential analysis of a scientific text through semiotics reveals the essence of the elaborate communicative structure behind the methods of laboratory investigation. In essence, I pick one event in molecular biological discovery, the association between regulation of gene expression and dynamic chromatin movement, and highlight how its arguments are articulated through the use of narrative and semiotic methods.

Porter-Abbott, Kay and Hoffmeyer have presented ways of conceiving of scientific communication narratively and semiotically. Porter-Abbott offers a portrait of the theory of evolution as narrative. Kay gives a semiotic representation of the elucidation of the genetic code and Hoffmeyer similarly addresses the role of semiotics in the scientific communicative apparatus. In these assessments, scientific texts are seen as rhetorical operations designed to persuade the reader of the veracity of the arguments. My elaboration consists of ascribing the differential nature of language to the arguments, as observations, results and conclusions. The semiotic dissection of the review articles emphasizes that the argument is contingent upon the binaries that the scientific terms set up. For example, the concept of nucleosomal perturbation as affecting gene expression is not demonstrable without evoking the notion of its binary term, static nucleosomal positioning. I stress this article because, in addition to providing a pivotal moment in molecular biological discovery, it offers a potent example of the essential movements in the process of constructing a scientific argument in the rapidly evolving field of transcriptional biology.

The originality of this analysis lies in bringing together these specific interpretations of scientific discourse, including scientific communication and the processes of laboratory investigation, i.e., the materials and methods that lead to molecular biological discovery. The differential nature of language, whereby knowledge is not produced without evoking its binary, and leads to distinction between what is accorded the status of "scientific fact" and what is not constituted as scientific fact.

In other words, to dissect scientific statements through narrative analysis, as developed by Barthes and Greimas, would lead to a greater understanding of the how scientists operate. In a sense, to promote greater self-awareness among molecular biologists is to subject the scientific text, which intricately sets up the binaries and oppositions through an interpretation of the terms its sets up, to semiotic analysis

In conclusion, my method is one in which I apply the tools of semiotic analysis to scientific inquiry and communication. In considering the epistemology of molecular biology, I ascribe how meaning is generated in a scientific text.

As articulated by Barthes, connotative signs are present in every text and one should be a producer of the text. I argue that each reader of a scientific text is a producer of the text. The elementary structure of signification, as evident in the Greimasian semiotic square, ascribes the differential meaning operating between the neutral terms of the scientific text.

I end this text with a note that my analysis is predicated upon assumptions that scientific communication is rhetorically constituted, and that semiotic and narrative apparati are useful in assessing its intellectual framework. To be aware of French narrative theory and the broad range of textual tools that I have laid out illuminates the processes of scientific discovery in molecular biology.

Appendix A

Figures 11 through 16 are reproduced with permission from *Genes VII,* Benjamin Lewin Copyright(2000) Oxford University Press.

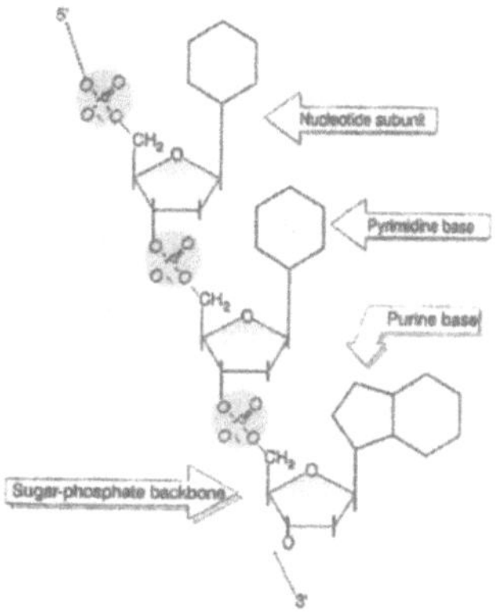

Figure 11 Chemical Structure of the DNA Molecule

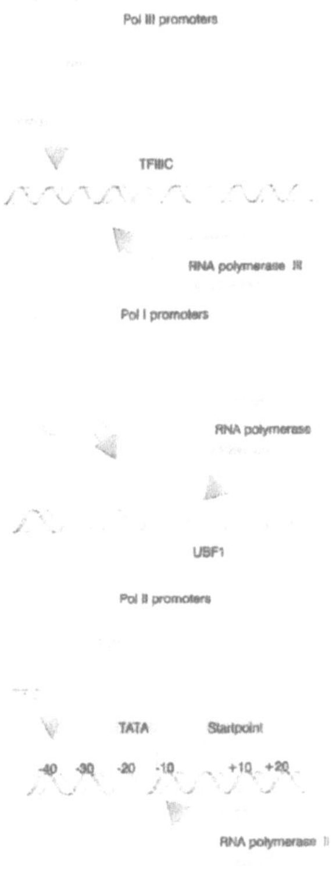

Figure 12 RNA polymerases are positioned at all promoters by TBP

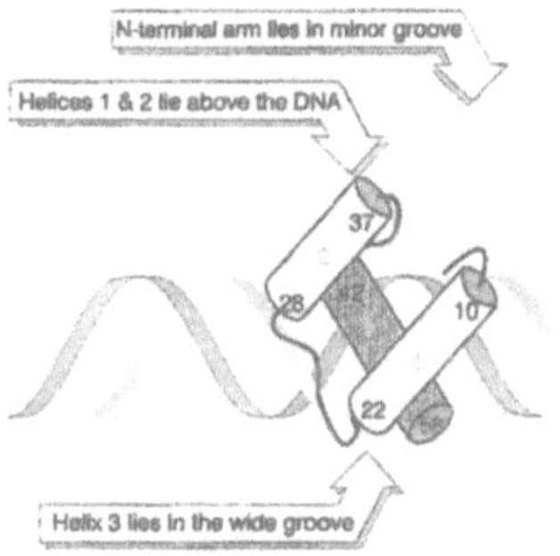

Figure 13 Helix 3 of the homeodomain binds in the major groove of the DNA with the helices 1 and 2 lying outside of the double helix

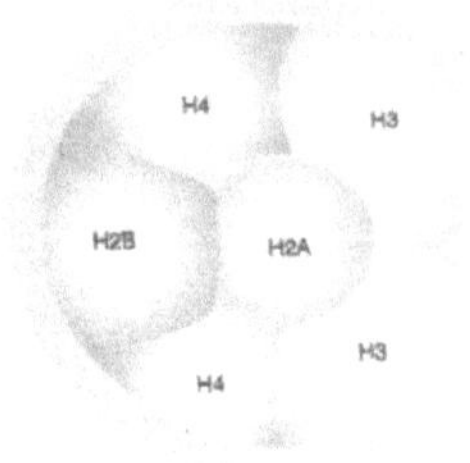

Figure 14 A symmetrical model of the nucleosome showing the histone octamer

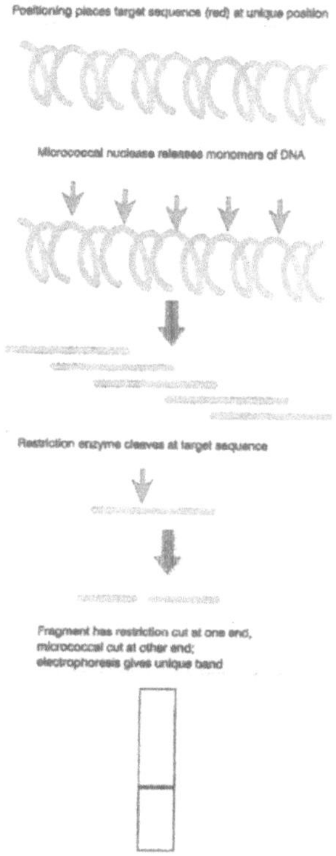

Figure 15 Nucleosomal positioning places restriction sites at unique positions relative to the linker sites cleaved by micrococcal nuclease

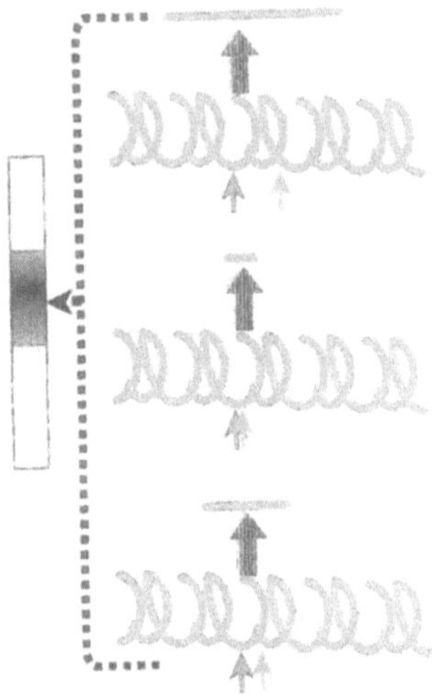

Figure 16 In the absence of nucleosomal positioning, a restriction site lies at all possible locations in different copies of the genome. Fragments of all sizes are produced when a restriction enzyme cuts at the target site and micrococcal nuclease cuts at the junctions between nucleosomes.

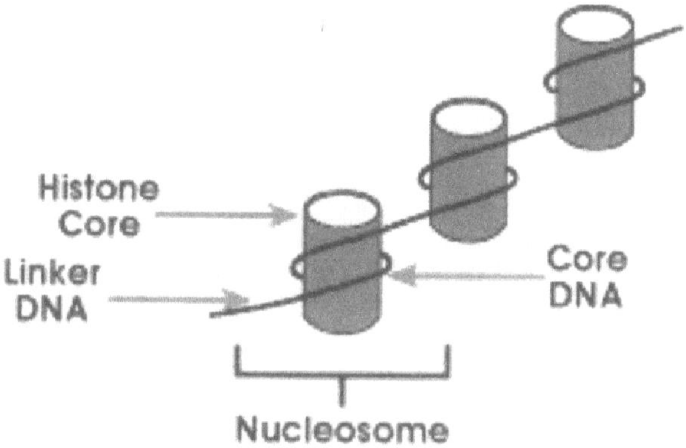

Figure 17 Nucleosomes: Histone and DNA

http://cnx.org/content/m11320/latest/histone.gif

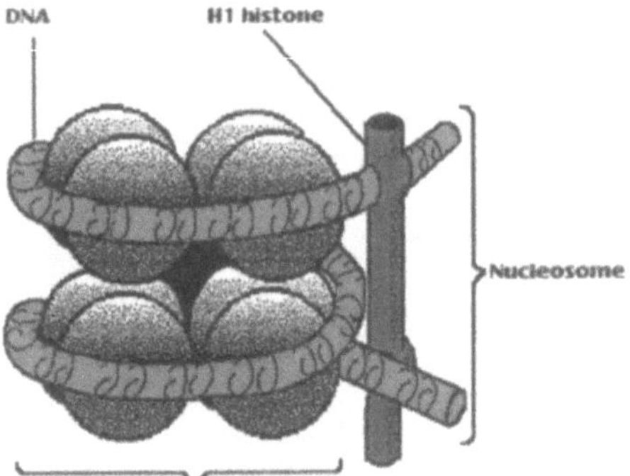

Figure 18 Packaging of Nucleosomes

http://www.accessexcellence.org/RC/VL/GG/images.nucleosome/gif

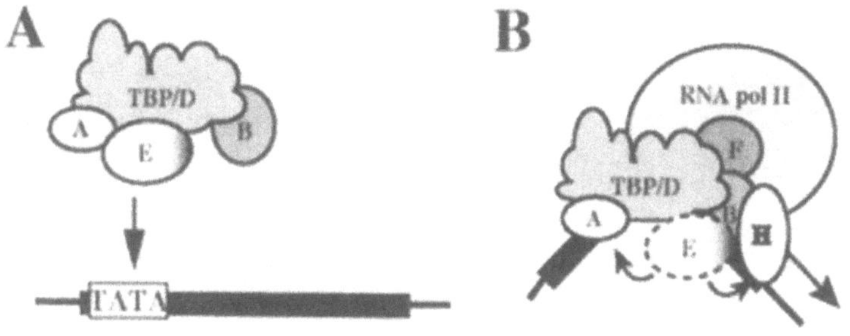

Figure 19 The TATA-Binding Protein binds to the TATA promoter

Yokomori, K., Verruzer, C.P., Tjian R. "An interplay between TATA-box binding protein and transcription factors IIE and IIA modulates DNA binding and transcription", *Proceedings National Academy of Science*, Copyright(1998), National Academy of Sciences

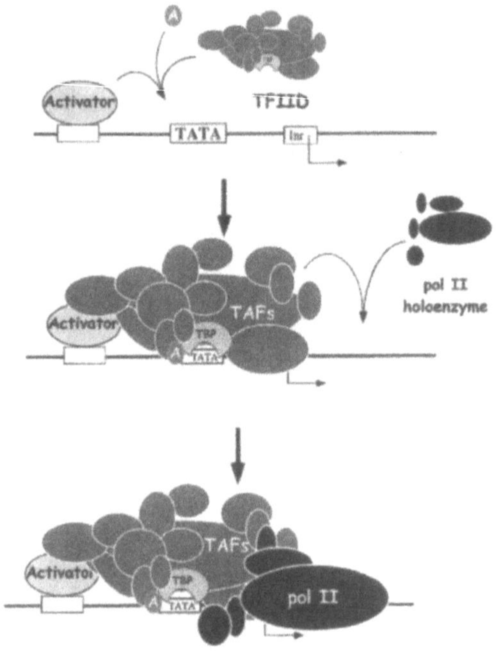

Figure 20 The transcription assembly complex

Reprinted from *Gene*, 255, B. Franklin Pugh, "Control of Gene Expression through regulation of the TATA-binding Protein," pp. 1-14, Copyright(2000), with permission from Elsevier

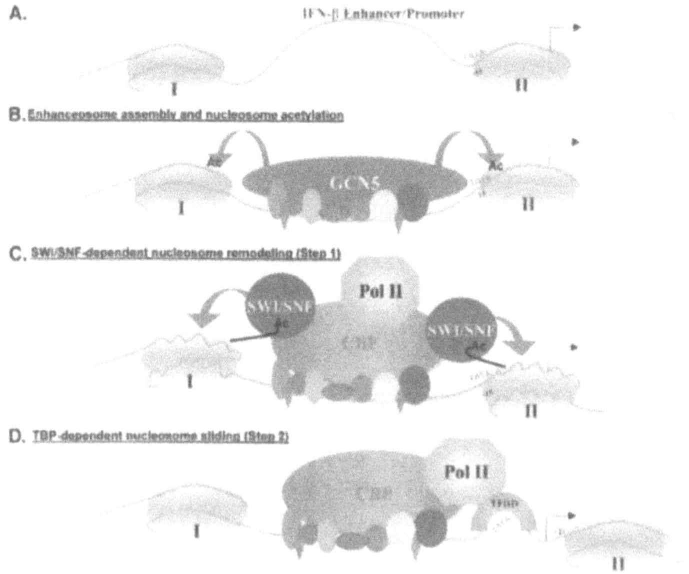

Figure 21 Acetylation of histones to activate transcription

Reprinted from *Cell*, 106, Stavros Lomvardos and Dimitris Thanos, "Nucleosomal sliding via TBP DNA Binding in Vivo," pp. 685-696, Copyright(2001), with permission from Elsevier

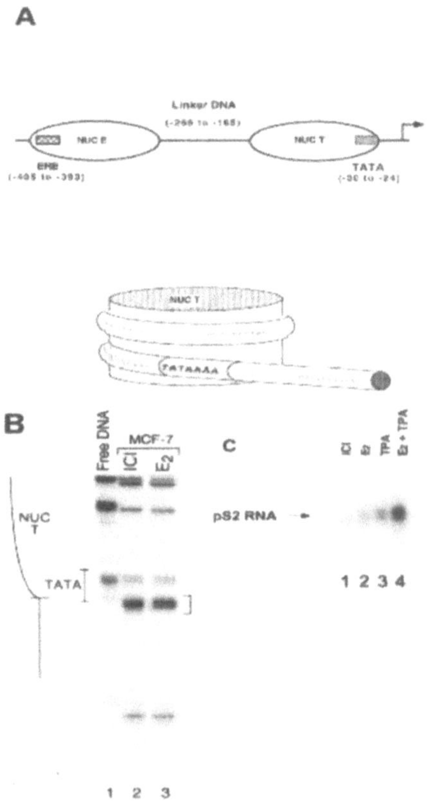

Figure 22 Nucleosomes are naturally positioned to allow for active transcription at the TATA promoter; chromatin remains passive

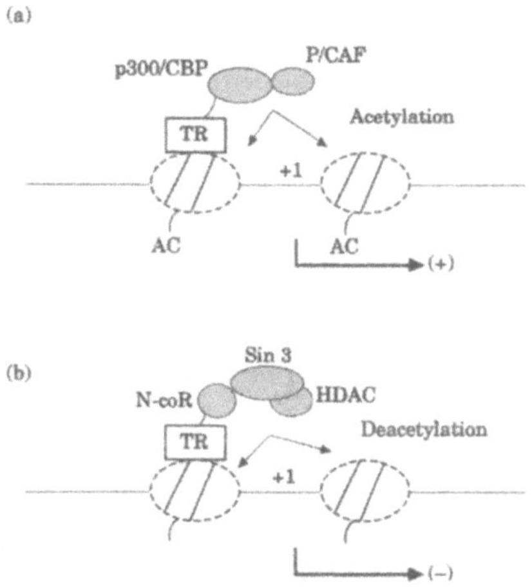

Figure 23 Nucleosomal structure is modified via recruitment of histone acetyltransferases (transcription) and histone deacetylases (repression)

Reprinted from *Molecular and Cellular Cardiology*, 30, W-Y. Huang and C-C Liew, "Chromatin Structure and Cardiac Gene Expression," pp. 1673-1681, Copyright(1998), with permission from Elsevier

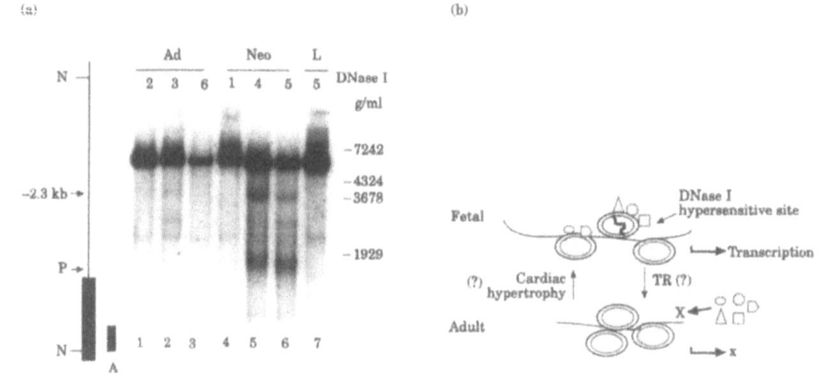

Figure 24 Nucleosomes actively move to allow transcription to take place

Appendix B

MOLECULAR AND CELLULAR BIOLOGY,
0270-7306/01/$04.0010
DOI: 10.1128/MCB.21.4.1404–1415.2001
Feb. 2001, p. 1404–1415 Vol. 21, No. 4

Binding of TATA Binding Protein to a Naturally Positioned Nucleosome Is Facilitated by Histone Acetylation

GERALD F. SEWACK, THOMAS W. ELLIS, AND ULLA HANSEN*

Department of Biology, Boston University, Boston, Massachusetts 02215

Received 31 October 2000/Accepted 21 November 2000

The TATA sequence of the human, estrogen-responsive pS2 promoter is complexed in vivo with a rotationally and translationally positioned nucleosome (NUC T). Using a chromatin immunoprecipitation assay, we demonstrate that TATA binding protein (TBP) does not detectably interact with this genomic binding site in MCF-7 cells in the absence of transcriptional stimuli. Estrogen stimulation of these cells results in hyperacetylation of both histones H3 and H4 within the pS2 chromatin encompassing NUC T and the TATA sequence. Concurrently, TBP becomes associated with the pS2 promoter region. The relationship between histone hyperacetylation and the binding of TBP was assayed in vitro using an in vivo-assembled nucleosomal array over the pS2 promoter. With chromatin in its basal state, the binding of TBP to the pS2 TATA sequence at the edge of NUC T was severely restricted, consistent with our in vivo data. Acetylation of the core histones facilitated the binding of TBP to this nucleosomal TATA sequence. Therefore, we demonstrate that one specific, functional consequence of induced histone acetylation at a native promoter is the alleviation of nucleosomemediated repression of the binding of TBP. Our data support a fundamental role for histone acetylation at genomic promoters in transcriptional activation by nuclear receptors and provide a general mechanism for rapid and reversible transcriptional activation from a chromatin template.

Transcription by RNA polymerase II (RNA pol II) from most eukaryotic promoters requires the evolutionarily conserved interaction of TATA binding protein (TBP) with a TATA sequence (42) in a chromatin context. Transcriptional activity, at least in yeast, correlates strongly with the degree of TBP occupancy of TATA box elements (30, 32). However, TBP can be severely inhibited from binding DNA sites located within unaltered mononucleosomes (17, 24).

* Corresponding author. Mailing address: Department of Biology, Boston University, 5 Cummington St., Boston, MA 02215. Phone: (617) 353-8730. Fax: (617) 353-8734. E-mail: uhansen@bu.edu.

Positioning of a nucleosome over the TATA region appears to be a common-mechanism for repressing basal transcription. In yeast, the TATA sequences of the inactive PHO5 (2), ADH2 (54), GAL80 (34, 35), and CHA1 (37) promoters are all contained within positioned nucleosomes that are disrupted during induction-dependent chromatin rearrangements. Similar mechanisms have been observed in other organisms. Both the repressed human immunodeficiency virus type 1 promoter in unstimulated human T cells (53) and the inactive beta phaseolin gene in vegetative tobacco tissues (31) contain nucleosomes that occlude their respective TATA sequences and are subsequently disrupted concomitant with transcriptional activation.

Of the multiple modifications required for conversion of chromatin from an inactive to active state, one consistent feature of active chromatin is the highly acetylated state of the core histones in the nucleosomes (13, 22, 52). Core histone acetylation influences both the interaction of specific proteins with nucleosomal DNA and the activation of gene expression (reviewed in reference 57). A broad range of transcriptional regulatory proteins possess intrinsic histone acetylase and deacetylase activity (for review, see references 49, 56, and 61). Included among the proteins that possess histone acetyltransferase (HAT) activity are several nuclear receptor coactivatorsthat interact directly with the estrogen receptor (ER), including SRC-1/NCoA-1, ACTR/RAC3/(P/CIP), TIF2/GRIP1/NCoA-2, and CBP (also called p300), as well as the CBPassociated factor (P/CAF) (11). Of these coactivators, the HAT activity of p300 has recently been demonstrated to be critical for hormone-induced histone H4 hyperacetylation on the pS2 promoter (9). The coactivators CBP and pCAF can also acetylate nonhistone proteins, including transcription factors p53, E2F1, ELKF, GATA 1, TFIIF, and TFIIEb and the nuclear receptor coactivator ACTR (reviewed in reference 29). While the crucial role of the acetylase activity of the nuclear receptor coactivators in hormone-induced gene regulation has-been demonstrated (9), the functional consequences of acetylation of core histones remain unresolved.

Given the complexity of the structural parameters governing binding of transcription factors to nucleosomal templates, an understanding of transcriptional activation at a natural promoter requires knowledge of the native chromatin structure. To this end, we have previously mapped the chromatin structure of the human, estrogen-responsive pS2 promoter within the context of its normal genomic location in human mammary epithelial cells (48). The TATA box at 230 to 224 of the pS2 promoter is situated at the 39 edge of a rotationally and translationally positioned nucleosome, NUC T (at nucleotides 223 to 2165). NUC T remains stably positioned even upon the transcriptional induction of the gene in vivo (48). This contrasts with the previously discussed nucleosomal disruptions over other transcriptionally active TATA sequences (2, 31, 34, 35, 37, 53, 54) and suggests an alternative mechanism of alleviating nucleosomal repression of the binding of TBP. The pS2 promoter can be activated through several estrogen receptordependent pathways, including direct binding of steroid hormone to the receptor (6) and phosphorylation of the receptor resulting from activation of the protein kinase C (PKC) pathway (39) or the Ras mitogen-activated protein kinase cascade of the growth factor signaling pathway (28). Here we examine how induction by estradiol and the PKC pathway results in transcriptional activation of the pS2 gene on a chromatin template. First, we establish that the binding of TBP is intrinsically linked to ER-dependent transcriptional activation of the pS2 promoter. Second, we demonstrate that the core histones H3 and H4 of the nucleosomal pS2 promoter are differentially hyperacetylated in response to these transcriptional stimuli. And most significantly, we demonstrate that hyperacetylation of core histones facilitates the binding of the basal transcription factor, TBP, to the TATA sequence within the pS2 chromatin.

MATERIALS AND METHODS

Cell cultures. CMT cells (an African green monkey kidney cell line that stably produces simian virus 40 [SV40] T antigen) and the human breast adenocarcinoma cell line MCF-7 were maintained in Dulbecco's modified Eagle's medium with 10% fetal calf serum in a 5% CO2 atmosphere at 37°C. Proliferating MCF-7 cells were grown in phenol red-free EX-CELL 320, serum-free media (JRH Biosciences) to eliminate any estrogenic activity from the media for 48 h prior to experimental treatment and chromatin immunoprecipitation (ChIP). In addition, the antiestrogen ICI 182,780 (Tocris) was added at 2 mM for this 48-h period to the unstimulated samples to fully block ER activation of the pS2 promoter. Cells were stimulated with 200 nM 17b-estradiol (Sigma) and 100 ng of tetradecanoyl phorbol acetate (TPA) (Sigma)/ml for 3 h prior to harvesting for ChIP assays.

ChIP and in vivo TBP binding assay. Approximately 3 3 108 MCF-7 cells were used for each experimental group. ChIPs were performed essentially as previously described (5; as adapted for mammalian cell culture per instructions of Upstate Biotechnology). After the designated treatments, cells were fixed by the direct addition of formaldehyde to the culture medium to a final concentration of 1% and incubation for 10 min at 37°C. Cells were washed with ice-coldphosphate-buffered saline containing protease inhibitors (1 mM phenylmethylsulfonyl fluoride, 1 mg of aprotinin/ml, and 1 mg of pepstatin A/ml) and were lysed with 1.2 ml of sodium dodecyl sulfate (SDS) lysis buffer (1% SDS, 10 mM EDTA, 50 mM Tris-HCl [pH 8.1]) containing the same protease inhibitors. This lysate was sonicated three times for 30 s using a Branson Sonifer 250 at 20% constant maximal power, resulting in an average DNA length of approximately 500 bp, as determined by electrophoresis through a 1% agarose–Tris-acetate-EDTA gel and ethidium bromide staining. After clarification, the supernatant fraction was diluted 10-fold in ChIP dilution buffer (0.01% SDS, 1.1% Triton X-100, 1.2 mM EDTA, 16.7 mM Tris-HCl [pH 8.1], 167 mM NaCl) to a final volume of 12 ml. Protein levels of the 12-ml samples were quantitated with a detergent-compatible protein assay kit (Bio-Rad) per manufacturer's directions, and protein concentrations were equalized. Immunoprecipitations were performed by incubating 2 ml of sample at 4°C overnight with saturating amounts of antibodies: 10 ml of anti-acetyl-histone H4 rabbit antiserum or 10 mg of antiacetyl-histone H3 rabbit polyclonal immunoglobulin G antibody (Upstate Biotechnology). Isolation of the immunocomplexes and further purification of the coprecipitated DNA were conducted as described previously (5). Specific sequences in the immunoprecipitates were detected by PCR, under conditions optimized for each primer set in which the yield of product was within the linear range relative to input DNA (data not shown). Identical immunoprecipitates were assayed for levels of the pS2 proximal promoter sequence, using primer set TE (59-ATGGGC-TTCATGAGCTCC and 59-GCGACCCCGAGTCAGG), and for the control sequence, beta interferon proximal promoter sequence, using primer set F (41). Immunoprecipitations and PCRs were performed in duplicate in each experiment, and multiple independent experiments were performed to ensure reproducibility. Measurement of the in vivo interaction of TBP with the pS2 proximal promoter was performed by modifying the ChIP protocol. Instead of anti-acetylated histone antibodies, 25 mg of anti-TBP antibodies (Santa Cruz Biotechnology) was used to immunoprecipitate cross-linked protein-DNA complexes followed by quantitative PCR analysis of the purified MCF-7 DNA fragments.

MC and core MC isolation. CMT cells were infected with the recombinant SV40/pS2 promoter virus SVSpS2, and minichromosomes (MCs) were isolated and purified as described previously (21). Where indicated, CMT cells were treated with 400 nM trichostatin A (TSA) (Waco Pure Chemical Inc.) 3 h prior to MC isolation (21). Core MCs were isolated after incubation in 0.5 M NaCl as described previously (16), except that bovine serum albumin (BSA) (Worthington Biochemicals) was added to a concentration of 0.1 mg/ml before the final dialysis step at 4°C against 30 mM HEPES (pH 7.5), 1 mMEDTA, 0.1% Nonidet P-40 (ICN Biomedicals Inc.), 1 mM 2-mercaptoethanol, and 15% sucrose.

TBP-DNA binding reactions. Human TBP, a glutathione transferase-tagged fusion protein (Santa Cruz Biotechnology), was added to either SVSpS2 viral DNA or to the indicated MCs. Human TFIIA, added to MCs in addition to TBP, was overexpressed in bacteria and was purified by affinity chromatography as described previously (14). The binding reactions for TBP with free DNA contained 55 mM KCl, 10 mM NaCl, 6 mM MgC12, 0.1 mg of BSA/ml, 20 mM HEPES (pH 7.5), 20% glycerol, and 10 mM dithiothreitol. The binding reactions for TBP on MCs and core MCs contained 47 mM KCl, 10 mM NaCl, 3 mM MgC12, 0.1 mg of BSA/ml, 20 mM HEPES (pH 7.5), 10% glycerol, 3% sucrose, and 10 mM dithiothreitol. In addition, the core MC preparation added a final concentration of 0.2 mM EDTA, 0.2 mM 2-mercaptoethanol, and 0.02% NonidetP-40 to the reactions. All binding reactions were carried out at 30°C for 45 min in a final volume of 30 ml. Following this incubation with TBP, DNA or MCs were digested with DNase I (Worthington Biochemicals) at 0.005 or 0.05 mg/reaction mixture, respectively. The digestions were performed by adding 30 ml of DNase I in 150 mM sucrose, 80 mM KCl, 35 mM HEPES (pH 7.5), 5 mM K2HPO4, 5 mM MgC12, 0.1 mg of BSA/ml, and 2 mM CaC12 to the 30-ml binding reaction mixture for 2 min at 20°C. DNase I cleavage was terminated by theaddition of 1 mg of EcoRI-digested genomic DNA from CMT cells and 5 mg of yeast tRNA in 60 ml of 20 mMEDTA and 20 mM Tris (pH 7.8). This mixture was placed on ice for 10 min, followed by the addition of 0.2% cetyltrimethylammoniumbromide, and was incubated at 20°C for 10 min, andthe DNA was precipitated by centrifugation in a microcentrifuge for 10 min at 15,000 3 g. Thecetyltrimethylammonium bromide was removed by resuspension of the DNAtRNA pellet in 3 M sodium acetate, followed by ethanol precipitation. Ligationmediated PCR (LMPCR) of 25 ng of isolated DNA was performed with primer set NTV-T, as described previously (48). The end-labeled PCR product was visualized on preflashed Reflection NEF-495 autoradiography film (Dupont) following electrophoresis through a 6% polyacrylamide–7 M urea gel (1:15, bisacrylamide:acrylamide).

Micrococcal nuclease and Western blot analysis of chromatin templates. The nucleosomal structure of the pS2 promoter within MCs, core MCs, and genomic DNA in MCF-7 cells was monitored by micrococcal nuclease digestion and LMPCR as previously described (48). For Western blot analysis, 500 ng of MC DNA (containing approximately 500 ng of core histones) was electrophoresed through an SDS–12.5% polyacrylamide gel. Proteins were transferred onto a polyscreen polyvinylidene difluoride membrane (New England Nuclear), and the membranes were incubated with either anti-acetylated-histone H4 rabbit antiserum (1:2,000 dilution) or anti-acetylated-histone H3 rabbit polyclonal immunoglobulin G antibody (2 mg/ml) (Upstate Biotechnology) in phosphate-buffered saline containing 3% nonfat milk. Acetylated histone H4 and H3 blots were subsequently incubated with anti-rabbit horseradish peroxidase-conjugated antibody (1:3,000) (Bio-Rad) and anti-rabbit alkaline phosphatase-conjugated antibody (1:2,000), respectively. Immunoreactive species were visualized by chemiluminescence using either a Dupont NEN Renaissance kit and autoradiography or Amersham Pharmacia ECF Western blotting kit and Fluorimager (Molecular Dynamics) per manufacturers' directions. Additionally, duplicate samples not subjected to Western blotting were electrophoresed and visualized by silver staining, per directions of the manufacturer (Daiichi Co.).

RESULTS

ER activation induces hyperacetylation of histones and binding of TBP within the pS2 promoter. The basal transcription factor TBP can be severely inhibited from binding DNA sites located within unaltered mononucleosomes (17, 24). In the human estrogen-responsive pS2 promoter, the TATA sequence lies within a tightly positioned nucleosome (Fig. 1A) whose translational boundary is defined by nucleosome-dependent micrococcal nuclease-hypersensitive sites (48) (Fig. 1B, bracket). Strikingly, this nucleosome positioning remains unchanged upon transcriptional induction of the gene (48) (Fig. 1B, lanes 2 and 3). Given that the nucleosome-covering sequence is not disrupted upon induction, we examined whether NUC T might be altered instead by modifications that would allow the binding of TBP. In particular, the biological relevance of histone acetylation of the human pS2 proximal promoter to activation with ER-dependent stimuli was investigated in the human MCF-7 cell line, a breast cancer cell line that constitutively expresses the ER. Transcriptional activation of the pS2 gene both by estradiol and by the PKC activator TPA is dependent upon the ER, and induction by these pathways is synergistic (10, 26) (Fig. 1C, lanes 2 to 4). In our experiments, levels of pS2 mRNA in the MCF-7 cells were stimulated 5-fold by estradiol, 9.5-fold by TPA, and 29-fold by a combination of the two agents. Transcriptional induction by both stimuli is inhibited by pure antiestrogens (8).

The acetylation states of histones H3 and H4 within the pS2 promoter were analyzed under our experimental conditions using the ChIP as say. Sheared cellular chromatin was immunoprecipitated with antibodies raised against either a diacetylated synthetic peptide corresponding to residues 1 to 21 of histone H3 or a tetra-acetylated peptide corresponding to residues 2 to 19 of histone H4. As a control, chromatin was also mock immunoprecipitated in the absence of antibody. The DNA from these precipitates was initially analyzed by quantitative PCR utilizing primers that flank sequences containing NUC T. The background level of recovered DNA in the mock immunoprecipitates was identical, whether derived from cells treated with ICI, estradiol, or TPA (data not shown).

With the ChIP assays, significant changes in modification states were demonstrated upon induction of gene expression. With antibodies reactive against acetylated histone H3, we observed a significant enhancement in immunoprecipitation of the pS2 promoter region in response to treatment with estrogen, as compared to treatment with the antiestrogen ICI, and an even greater enhancement upon treatment with both estradiol and TPA (Fig. 2A and C). A somewhat distinct result was obtained with antibodies reactive against acetylated histone H4. As compared to levels of the pS2 promoter DNA associated with acetylated histone H4 in the presence of ICI, equivalent increases were observed upon treatment with estradiol either in the presence or absence of TPA (Fig. 2A and C). We note that the amount of pS2 DNA immunoprecipitated from ICI-treated samples by the H4 antibodies was much higher than the amount immunoprecipitated by the H3 antibodies. This may reflect a higher basal level of histone H4 acetylation within the pS2 proximal promoter region in its uninduced state or may reflect recognition of all acetylated forms of histone H4 by the anti-acetylated-histone H4 antibodies (33), in contrast to recognition of only highly acetylated isoforms of histone H3 by the anti-acetylated-histone H3 antibodies (4).

In order to ensure that the enhanced histone acetylation associated with the pS2 promoter did not reflect global nucleosome modifications, we examined the acetylation state of the chromatin associated with the beta interferon proximal promoter region, encompassing its TATA sequence. Analysis of the identical chromatin immunoprecipitates in three independent experiments revealed no statistical change in the acetylationstate of histones H3 or H4 associated with this control promoter upon addition of estradiol and TPA (Fig. 2B and C). This correlates with nondetectable levels of beta interferon mRNA in samples isolated from MCF-7 cells treated with ICI or estradiol and TPA, as analyzed by Northern blotting (data not shown).

In these assays, the levels of histone acetylation were determined following 3 h of exposure of the cells to the hormone. However, increases in acetylation of both histones H3 and H4 were observed as early as 15 min after treatment with estradiol (data not shown). These data are generally consistent with the recent findings of Chen and colleagues (9), although they differ in details (see Discussion).

The influence of transcriptional stimuli on the binding of the critical downstream general transcription factor, TBP, to the genomic pS2 proximal promoter was also examined using an immunoprecipitation assay. The experimental design was identical to the ChIP assay, except for the use of anti-TBP antibodies. The basal level of recovered DNA after a mock immunopre cipitation protocol was the same as the level of DNA recovered in anti-TBP immunoprecipitates from extracts of ICI-treated cells (data not shown). Thus, in the presence of the pure antiestrogen ICI 182,780, TBP did not detectably interact with this pS2 proximal promoter region, which includes the TATA sequence (Fig. 2D, lanes 5 and 7). This is a significant result, as TBP is sometimes associated with cellular promoters that remain in an inactive state (23).

Activation of ER by estradiol and TPA induced the binding of TBP to the pS2 promoter (Fig. 2D, compare lanes 5 and 7 with lanes 6 and 8). The level of binding of TBP in the induced state was fourfold higher than background levels. However, given that background levels in the absence of transcriptional-stimuli were the same with or without anti-TBP antibodies, the actual extent to which transcriptional activation of the genomic pS2 promoter by ER promotes binding of TBP is undoubtedly much greater.

The nucleosome naturally positioned over the pS2 TATA sequence severely restricts accessibility to TBP binding. Given that core histones complexed with the genomic pS2 promoter are specifically hyperacetylated in response to transcriptional stimuli and that these same stimuli lead to increased binding of TBP, we investigated a possible functional connection at this promoter between histone acetylation and the binding of this basal transcription factor. Assembly of the preinitiation complex at most RNA pol II promoters is critically dependent on the binding of TBP with its associated factors to a TATA sequence, as RNA pol II cannot directly recognize target initiation sites (reviewed in reference 7). The binding of TBP to the TATA box within the positioned nucleosome of the human pS2 promoter was studied using a minichromosome model system. Previously, it was demonstrated that nucleotides 21100 to 110 of the pS2 promoter are sufficient to establish the positioning of NUC T, as this region of the promoter, when placed in the context of a recombinant SV40 viral genome, generates in vivo-assembled chromatin templates with structural features virtually identical to those of the genomic pS2 promoter (48). SV40 MCs are comprised of an array of the normal cellular nucleosomes, including histone H1, complexed with the viral genome. Nonhistone proteins are also bound to the MCs (for review, see references 12 and 15).

The basic characteristics of the TBP footprinting pattern and its binding avidity to the pS2 TATA sequence were initially defined on deproteinized, recombinant viral DNA (Fig. 3A). Due to the sequence specificity of DNase I, there was little cleavage of the free DNA directly within the TATA sequences (Fig. 3A, lane 1). However, a characteristic footprint was nonetheless evident upon addition of increasing amounts of TBP. The binding of TBP resulted in protection against cleavage at the 59 edge of the TATA box from nucleotides 235 to 229 (Fig. 3A, lanes 5 to 7, bracket). DNase I footprints of TBP extending 59 of the TATA sequence are commonly observed, such as on the adenovirus major late promoter TATA sequence, where the protected footprint extends to nucleotide 236 (62). The interaction of TBP with the TATA sequence also caused several base pairs immediately 59 of the footprinted region (nucleotides 237 and 238) to become hypersensitive to DNase I cleavage. DNase I-hypersensitive sites have also been observed on the adenovirus major late and human heat shock protein 70 promoters, 59 of the regions footprinted by TFIID (38). The apparent affinity of TBP for the free pS2 TATA sequence is quite weak. In these experiments, the half-maximal concentration for binding is nearly 0.6mM TBP (0.2 mM active TBP). Thus, the apparent Kd is approximately 2 orders of magnitude lower than the reported affinity of TBP, 2.0 nM, for the adenovirus major late promoter TATA box (19, 25). This is due partly to differing sequences flanking the TATA boxes and also to binding conditions that are necessitated by the subsequent assays utilizing MCs (see below). In particular, BSA was present in our assays at concentrations that were previously shown to reduce the affinity of TBP for the adenovirus major late promoter TATA box by fivefold (25).

The general initiation factor TFIIA can enhance the binding of TBP to naked DNA when TBP binding conditions are suboptimal (25, 62), and it has therefore been included in previous studies of binding of TBP to reconstituted nucleosomal templates (17, 24). In the presence of TFIIA, the affinity of TBP for the naked pS2 TATA sequence under these binding conditions was increased but by less than 10-fold (compare Fig. 3A, lanes 5 and 6, with Fig. 3B, lane 4). On this DNA, the presence of TFIIA also enlarged the footprint significantly upstream of the TATA sequence (compare the bracketed areas in Fig. 3A and B).

When TBP alone was added in a parallel experiment to recombinant viral chromatin containing the pS2 promoter, nodetectable footprint was observed even at the highest concentrations of TBP (Fig. 3C, lane 3). Thus, the binding of TBP to the pS2 TATA sequence was strongly inhibited within the context of the positioned NUC T and surrounding chromatin, despite location of the TATA sequence at the extreme edge of NUC T (Fig. 1B). Upon inclusion of TFIIA, TBP binding to the nucleosomal pS2 TATA sequence in the MC context was still only minimally detectable at the highest concentrations of TBP (Fig. 3D and 4A, lanes 1 to 6). Although the evidence for an interaction between TBP and unmodified chromatin in the presence of TFIIA was subtle, often only the appearance of DNase I hypersensitivity at nucleotides 237 and 238, this result was reproducible. TFIIA was included in all subsequent experiments.

Enhanced binding of TBP to the nucleosomal pS2 TATA sequence in the context of highly acetylated chromatin. Given the extremely weak interaction of TBP with the nucleosomal pS2 TATA sequence, a role for histone acetylation on the binding of TBP to the nucleosomal pS2 promoter was tested, using highly acetylated chromatin templates. In order to isolate such chromatin, cells were infected with the recombinant SV40 virus containing the pS2 promoter and were treated subsequently with TSA. By inhibiting histone deacetylase activity, TSA causes accumulation in vivo of highly acetylated corehistones (63). Western blotting analyses confirmed the higher levels of acetylation of histones H3 and H4 within MCs isolated from cells incubated with TSA (Fig. 4D).

Incubation of TBP with this highly acetylated chromatin led to stable, highly specific binding at the pS2 TATA sequence (Fig. 4A, lanes 7 to 13, footprint over 59 half of TATA box, nucleotides 230 to 234, as marked by bracket). The binding of TBP was dramatically enhanced as compared to its binding to MCs isolated from untreated cells. The differences in binding to the two chromatin preparations were quantitated both as protection from DNase I cleavage of individual bases (Fig. 4B) and as protection of the entire footprinted region relative to a region outside of the binding area (3) (Fig. 4C). At a TBP concentration (130 nM) resulting in 50% occupancy of the pS2 TATA sequence within the TSA-treated MCs (Fig. 4A, lane 9; B, blue line, TSA-treated chromatin; and C, 0.13 mM), binding to the same sequence within the unmodified MCs was undetectable (Fig. 4A, lane 3; B, blue line, normal chromatin; and C, 0.13 mM). In fact, acetylation of the MCs and the presence of TFIIA permitted levels of binding of TBP to the nucleosomal pS2 TATA sequence roughly comparable to those for binding of pure TBP to free DNA (compare Fig. 3A and 4A, noting the correction for the specific concentration of active TBP in each experiment, as indicated in the Fig. 3A legend).

Acetylation specifically of the core histones facilitates binding of TBP to NUC T. MCs assembled in vivo contain many chromatin components that may be acetylated in addition to the core histones (for example, see reference 9). In order to test definitively whether acetylation specifically of the corehistones on the MCs facilitated the binding of TBP, we analyzed MCs containing only the core histones. Incubation of the native MC preparations with 0.5 M NaCl prior to purification by sucrose gradient sedimentation dissociates histone H1, along with all nonhistone proteins normally associated withnative MCs, while leaving the core nucleosomes intact (16). Such core MCs were prepared from both normal and highly acetylated preparations of MCs. It was critical to verify that the in vivo translational positioning of NUC T was retained following the high-salt treatment. Therefore, both MCs and core MCs were analyzed for positioned nucleosomes by digestion with micrococcal nuclease. Nucleosome-dependent micrococcal nuclease-hypersensitive cleavages at nucleotide positions 223 and 224 are indicative of the edge of NUC T in the genomic pS2 promoter (48) (Fig. 1B and 5B, lane 7). These hypersensitive cleavages were evident in all purified MC and core MC samples of the recombinant SV40/pS2 promoter virus SVSpS2 (Fig. 5B, lanes 3 to 6) but not in the deproteinated viral DNA samples (Fig. 1B, lane 1, and 5B, lanes 1 to 2), proving that the nucleosomes remained positioned over the appropriate sequences under all conditions.

Stable binding of TBP was readily observed on the core nucleosomal templates prepared from highly acetylated MCs (Fig. 5A, bracket, lanes 9 to 17), as evidenced by complete protection from DNase I cleavage within the 59 region of the TATA sequence. The footprint in this experiment encompassed protection of one major cleavage site near the 59 end of the TATA sequence, as well as two minor cleavage sites in the center and on the 39 end. In contrast, very limited protection from DNase I cleavage was observed within this region on normal core MCs when titrated with increasing amounts of TBP (Fig. 5A, lanes 1 to 8). These data indicate that acetylation of core histones is sufficient to facilitate the binding of TBP to its nucleosomal site in the pS2 promoter.

At high concentrations of TBP, enhanced sensitivity to DNase I cleavage upstream of the TATA sequence was obtained on both normal and highly acetylated core MCs (Fig. 5A, lanes 6 to 8 and 14 to 16, region above bracket). Notably, on the acetylated core MCs, significant amounts of hypersensitivity occurred at TBP concentrations higher than those sufficient to observe DNase protection. While a DNase I footprint results from continuous protection of a specific DNA sequence from nuclease throughout the entire digestion period, thereby signifying a stable interaction between a protein and DNA, hypersensitivity to cleavage by DNase I can capture a transient interaction between a protein and DNA. Thus, we infer that transient binding of TBP to the normal core MCs did occur, with the dissociation rate being too fast to permit a stable interaction at the TATA sequence.

DISCUSSION

Many reports, beginning with in vitro transcription studies of the adenovirus major late promoter, have demonstrated that when a TATA sequence is placed within a chromatin context, competition between the binding of TBP and of the core histones leads to repression of transcription by nucleosomes (36, 58, 59). Similarly, we demonstrate that on a native promoter, a highly positioned nucleosome, NUC T, represses binding of TBP to the pS2 TATA sequence in vitro within an array of positioned nucleosomes. Furthermore, we provide evidence that repression by chromatin occurs at this promoter in vivo, as the binding of TBP to the genomic pS2 TATA sequence is repressed in the absence of activated ER (Fig. 2D).

How such inhibition can be overcome is critical for understanding transcriptional activation of this type of promoter. Our data are the first to directly show that acetylation of the core histones is sufficient to permit a stable interaction between TBP and a nucleosomal TATA sequence at a native promoter. In a previous report, the presence of the amino-terminal tails of the core histones in a mononucleosome was shown to decrease the accessibility of a nucleosomal TATA box (17). These data were interpreted to suggest that histone acetylation would enhance the binding of TBP to a nucleosomal TATA sequence. However, histone acetylation and proteolytic tail removal have recently been shown to have distinct effects on nucleosomes, with histone acetylation unexpectedly increasing nucleosome stability (55). Thus, effects of acetylation must be the result of defined changes in the specific structures and conformations adopted by the histone tails rather than a general charge neutralization effect (20). In our study, the binding of TBP was not only investigated with acetylated histones but also on a naturally positioned nucleosomal array in a native promoter. This provides biochemical evidence of histone acetylation facilitating the binding of TBP in a biologically relevant context.

ER and histone acetylation. The facilitated binding of TBP is a critical step in our proposed model of transcriptional activation of the genomic pS2 promoter by active ER. We suggest that binding of ER to the nucleosomal pS2 templaterecruits coactivators, which results in targeted histone acetylation of adjacent nucleosomes, including NUC T encompassing the TATA sequence. Histone acetylation of NUC T then facilitates the binding of TBP (as part of the TFIID complex) and subsequent transcription of the gene. This model proposes that the regulated accessibility of the TATA sequence is a necessary but perhaps not sufficient step for transcriptional activation. Multiple regulatory steps seem to be involved in activation of the pS2 promoter by ER. This is demonstrated by the observation that elevated levels of core histone acetylation, resulting from treatment of MCF-7 cells with the deacetylase inhibitor TSA, do not constitutively activate transcription from this promoter in vivo (43; G. F. Sewack and U. Hansen, unpublished observations). In vitro transcription studies have also demonstrated the complexity of ligand-dependent activation by ER. In a highly purified human transcription system, the TBP-associated factors (TAFIIs), the other components of TFIID, were strictly required for activation both of a synthetic DNA promoter and of a chromatin template (60). In fact, direct protein-protein interactions between ER and a variety of potential targets in the basal transcription machinery have been well documented (27, 45, 46). Therefore, transcriptional activation of the genomic pS2 promoter by ER is likely to involve multiple mechanisms, including both direct factor recruitment via protein-protein interactions and directed chromatin modification (discussed in detail below). The requirement for both functions of classical activation domains has recently been demonstrated, using synthetic yeast promoters in which the TATA sequence was assembled into a nucleosome. In this system, an artificially recruited TFIID, which essentially mimics the direct recruitment of TBP to a promoter, was insufficient to activate promoters with nucleosomal TATA elements (44).

The importance of histone acetylation in activation of transcription from estrogen-responsive promoters is supported by a variety of data. Histone acetylation is not required, however, for interaction of ER with the chromatin template, as ER binds effectively in vitro to the estrogen response element (2405 to

2393) within nucleosome E of the chromatin-assembled pS2 promoter (Sewack and Hansen, unpublished). Therefore, the critical question in this regard is how binding of ER activates the promoter in its native chromatin context to facilitate binding of subsequent transcription factors. ER contains two activation domains, AF-1 and AF-2. The activity of the ligand-independent, N-terminal AF-1 is stimulated by phosphorylation of serine residues (1, 18, 28). Such phosphorylation of the AF-1 domain in ERb by the mitogen-activated protein kinase pathway leads to the recruitment of the coactivator SRC-1 (51). Likewise, transcriptional induction by the conserved hormonedependent AF-2 domain is dependent on the recruitment of nuclear receptor coactivators, including, as possible candidates, SRC-1, TIF2/GRIP1, ACTR/AIB1/RAC3/pCIP, CBP, and pCAF, many of which possess HAT activity (9; for review, see references 11 and 50). Formation of a complex between activated ER and coactivators containing HAT activity can be an integral component of transcriptional activation by ER from a chromatin template, as was shown directly by the inability of HAT-negative coactivators to mediate estrogen-dependent transcription (9).

We show in this study that in vivo transcriptional stimulation results in hyperacetylation of histones H3 and H4 within the region of the pS2 promoter containing NUC T. Our data are in line with a recent study demonstrating enhanced acetylation of histones H3 and H4 within the proximal promoter regions of nuclear hormone receptor-responsive genes in response to hormone treatment (9), including both estrogen-responsive genes (genes for pS2, c-myc, EB-1, and cathepsin D) and retinoic acid receptor target genes (genes for p21 and CD38). For the cathepsin D and p21 promoters, enhanced acetylation was localized, in that the acetylation states of histones H3 and H4 located 3 to 4 kb upstream of the initiation sites were unchanged by hormone treatment (9). Furthermore, hyperacetylation of histone H4 was rapid and transient at several hormone-regulated promoters, correlating with RNA pol II association on the cathepsin D promoter and expression of this gene. These data suggested a general role for core histone acetylation in the rapid transcriptional activation of nuclear receptor target genes.

Localized hyperacetylation of histones H3 and H4, centered around the transcription start site of the beta interferon promoter, also correlated with transcriptional activation of this gene (41). Given that the highly localized region containing hyperacetylated core histones also contained the TATA sequence, facilitation of binding of TBP by histone acetylation may be a common mechanism of transcriptional activation at many RNA pol II promoters.

Even though the importance of histone acetylation in transcriptional activation is becoming increasingly clear, the direct effect of specific acetylation events remains unresolved. Specialized roles for individual coactivators in the transcriptional activation process at a given promoter are suggested by the different acetylation substrate specificities of the nuclear receptor coactivators (40,

47). Our study supports regulation by multiple coactivators, in that acetylation of histone H3 and histone H4 at the pS2 promoter was differentially regulated. In particular, whereas combined treatment of cells with TPA and estradiol resulted in consistently higher levels of histone H3 acetylation than with estradiol alone, enhancement of histone H4 acetylation was unchanged by the inclusion of TPA (Fig. 1C). The time courses of acetylation of histones H3 and H4 also appeared to differ. Histone H4 was transiently acetylated on the pS2 promoter (9). The time course of histone H4 acetylation correlated with association of ACTR, p300, and CBP coactivators with the promoter but not of ER, which remained bound to the pS2 proximal promoter for at least 6 h (9). In contrast, we demonstrate that highly acetylated histone H3 remained associated with the pS2 promoter even after 3 h of estradiol stimulation (Fig. 1C). By nuclear run-on assay, the pS2 gene remains maximally active at this time, in contrast to other estrogen-stimulated promoters (6). Taken together, these results suggest a role for histone H3 acetylation in the maintenance of a transcriptionally active promoter, even though histone H4 acetylation is transient. Histone H3 acetylation could be maintained by the activity of a different coactivator. We hypothesize that either acetylation event would facilitate the binding of TBP to its nucleosomal binding site in the promoter.

In conclusion, given the lack of disruption of highly positioned nucleosomes upon induction of the pS2 promoter and the reversibility of histone acetylation in this region, we speculate that acetylation of histones within NUC T, containing the pS2 TATA sequence, provides a rapid and reversible regulatory step characteristic of nuclear receptor-controlled transcription. And finally, because the binding of TBP to a nucleosomal TATA sequence represents a general regulatory step in RNA pol II transcription (reviewed in reference 7), we propose that the acetylation of core histones to facilit ate binding of TBP will provide a general mechanism of facilitating transcriptional activation at many TATA sequence-containing promoters.

ACKNOWLEDGMENTS

We thank Robert Roeder and Jeff Delong for TFIIA expression plasmids and Jeff Parvin for technical advice. We are also grateful to Tom Maniatis and Bhavin Parekh for the beta interferon PCR primers. The critical suggestions of Robert Kingston, Myles Brown, Fred Winston, Tony Imbalzano, Han-Fei Ding, Konstantin Ebralidse, and Geof Cooper are also greatly appreciated.This work was supported by American Cancer Society grants FRA-415, BE-231, and RPG-95-005 and by NIH training grant T32-CA09361.

REFERENCES

1. Ali, S., D. Metzger, J. M. Bornert, and P. Chambon. 1993. Modulation of transcriptional activation by ligand-dependent phosphorylation of the human oestrogen receptor A/B region. EMBO J. 12:1153–1160.
2. Almer, A., H. Rudolph, A. Hinnen, and W. Ho¨rz. 1986. Removal of positioned nucleosomes from the yeast PHO5 promoter upon PHO5 induction releases additional upstream activating DNA elements. EMBO J. 5:2689–2696.
3. Ausubel, F., R. Brent, R. Kingston, D. Moore, J. G. Seidman, J. Smith, and K. Struhl. 1993. DNA-protein interactions, p. 12.4. In K. Janssen (ed.), Current protocols in molecular biology. Wiley Interscience, New York, N.Y.
4. Braunstein, M., R. E. Sobel, C. D. Allis, B. M. Turner, and J. R. Broach. 1996. Efficient transcriptional silencing in Saccharomyces cerevisiae requires a heterochromatin histone acetylation pattern. Mol. Cell. Biol. 16:4349–4356.
5. Braunstein, M., A. B. Rose, S. G. Holmes, C. D. Allis, and J. R. Broach. 1993. Transcriptional silencing in yeast is associated with reduced nucleosome acetylation. Genes Dev. 7:592–604.
6. Brown, A. M. C., J.-M. Jeltsch, M. Roberts, and P. Chambon. 1984. Activation of pS2 gene transcription is a primary response to estrogen in the human breast cancer cell line MCF-7. Proc. Natl. Acad. Sci. USA 81:6344–6348.
7. Burley, S. K., and R. G. Roeder. 1996. Biochemistry and structural biology of transcription factor IID (TFIID). Annu. Rev. Biochem. 65:769–799.
8. Chalbos, D., A. Phillips, F. Galtier, and H. Rochefort. 1993. Synthetic antiestrogens modulate induction of pS2 and cathepsin-D messenger ribonucleic acid by growth factors and adenosine 39,59-monophosphate in MCF-7 cells. Endocrinology 133:571–576.
9. Chen, H., R. J. Lin, W. Xie, D. Wilpitz, and R. M. Evans. 1999. Regulation of hormone-induced histone hyperacetylation and gene activation via acetylation of an acetylase. Cell 98:675–686.
10. Cho, H., and B. S. Katzenellenbogen. 1993. Synergistic activation of estrogen receptor-mediated transcription by estradiol and protein kinase activators. Mol. Endocrinol. 7:441–452.
11. Collingwood, T. N., F. D. Urnov, and A. P. Wolffe. 1999. Nuclear receptors: coactivators, corepressors and chromatin remodeling in the control of transcription. J. Mol. Endocrinol. 23:255–275.
12. Cre´misi, C. 1979. Chromatin replication revealed by studies of animal cells and papovaviruses (simian virus 40 and polyoma virus). Microbiol. Rev. 43:297–319.
13. Csordas, A. 1990. On the biological role of histone acetylation. Biochem. J. 265:23–38.
14. DeJong, J., R. Bernstein, and R. G. Roeder. 1995. Human general transcription factor TFIIA: characterization of a cDNA encoding the small subunit and requirement for basal and activated transcription. Proc. Natl. Acad. Sci. USA 92:3313–3317.

15. DePamphilis, M. L., and P. M. Wassarman. 1982. In A. S. Kaplan (ed.), Organization and replication of viral DNA, p. 37–115. CRC Press, Boca Raton, Fla.
16. Ding, H.-F., M. Bustin, and U. Hansen. 1997. Alleviation of histone H1-mediated transcriptional repression and chromatin compaction by the acidic activation region in chromosomal protein HMG-14. Mol. Cell. Biol. 17: 5843–5855.
17. Godde, J. S., Y. Nakatani, and A. P. Wolffe. 1995. The amino-terminal tails of the core histones and the translational position of the TATA box determine TBP/TFIIA association with nucleosomal DNA. Nucleic Acids Res. 23:4557–4564.
18. Goff, P. L., M. M. Montano, D. J. Schodin, and B. S. Katzenellenbogen. 1994. Phosphorylation of the human estrogen receptor. J. Biol. Chem. 269:4458–4466.
19. Hahn, S., S. Buratowski, P. A. Sharp, and L. Guarente. 1989. Yeast TATA binding protein TFIID binds to TATA elements with both consensus and non-consensus DNA sequences. Proc. Natl. Acad. Sci. USA 86:5718–5722.
20. Hansen, J. C., C. Tse, and A. P. Wolffe. 1998. Structure and function of the core histone N-termini: more than meets the eye. Biochemistry 37:17637–17641.
21. Hansen, U. 1999. Transcriptional and structural analyses of isolated SV40 chromatin, p. 261–290. In P. B. Becker (ed.), Methods in molecular biology: chromatin protocols. Humana Press, Totowa, N.J.
22. Hebbes, T. R., A. W. Thorne, and C. Crane-Robinson. 1988. A direct link between core histone acetylation and transcriptionally active chromatin. EMBO J. 7:1395–1402.
23. Hoffmann, A., T. Oelgeschla¨ger, and R. G. Roeder. 1997. Considerations of transcriptional control mechanisms: do TFIID-core promoter complexes recapitulate nucleosome-like functions? Proc. Natl. Acad. Sci. USA 94:8928–8935.
24. Imbalzano, A. N., H. Kwon, M. R. Green, and R. E. Kingston. 1994. Facilitated binding of TATA-binding protein to nucleosomal DNA. Nature 370: 481–485.
25. Imbalzano, A. N., K. S. Zaret, and R. E. Kingston. 1994. Transcription factor (TF) IIB and TFIIA can independently increase the affinity of the TATA binding protein for DNA. J. Biol. Chem. 269:8280–8286.
26. Ingar-Trowbridge, D. M., M. Pimentel, M. G. Parker, J. A. McLachlan, and K. S. Korach. 1996. Peptide growth factor cross-talk with the estrogen receptor requires the A/B domain and occurs independently of protein kinase C or estradiol. Endocrinology 137:1735–1744.

27. Jacq, X., C. Brou, Y. Lutz, I. Davidson, P. Chambon, and L. Tora. 1994. Human TAFII30 is present in a distinct TFIID complex and is required for transcriptional activation by the estrogen receptor. Cell 79:107–117.
28. Kato, S., H. Endoh, Y. Masuhiro, T. Kitamoto, S. Uchiyama, H. Sasaki, S. Masushige, Y. Gotoh, E. Nishida, H. Kashime, D. Metzger, and P. Chambon. 1995. Activation of the estrogen receptor through phosphorylation by mitogen-activated protein kinase. Science 270:1491–1494.
29. Kouzarides, T. 2000. Acetylation: a regulatory modification to rival phosphorylation? EMBO J. 19:1176–1179.
30. Kuras, L., and K. Struhl. 1999. Binding of TBP to promoters in vivo is stimulated by activators and requires Pol II holoenzyme. Nature 399:609–613.
31. Li, G., S. P. Chandler, A. P. Wolffe, and T. C. Hall. 1998. Architectural specificity in chromatin structure at the TATA box in vivo: nucleosome displacement upon B-phaseolin gene activation. Proc. Natl. Acad. Sci. USA 95:4772–4777.
32. Li, X.-Y., A. Virasius, X. Zhu, and M. Green. 1999. Enhancement of TBP binding by activators and general transcription factors. Nature 399:605–609.
33. Lin, R., J. W. Leone, R. G. Cook, and C. D. Allis. 1989. Antibodies specific to acetylated histones document the existence of deposition- and transcription-related histone acetylation in Tetrahymena. J. Cell Biol. 108:1577–1588.
34. Lohr, D. 1993. Chromatin structure and regulation of the eukaryotic regulatory gene GAL80. Proc. Natl. Acad. Sci. USA 90:10628–10632.
35. Lohr, D., and J. Lopez. 1995. GAL4/GAL80-dependent nucleosome disruption/disposition on the upstream regions of the yeast GAL1–10 and GAL80 genes. J. Biol. Chem. 270:27671–27678.
36. Meisterernst, M., M. Horikoshi, and R. G. Roeder. 1990. Recombinant yeast TFIID, a general transcription factor, mediates activation by the gene-specific factor USF in a chromatin assembly assay. Proc. Natl. Acad. Sci. USA 87:9153–9157.
37. Moreira, J. M., and S. Holmberg. 1998. Nucleosome structure of the yeast-CHA1 promoter: analysis of activation-dependent chromatin remodeling of an RNA-polymerase-II-transcribed gene in TBP and RNA pol II mutants defective in vivo in response to acidic activators. EMBO J. 17:6028–6038.
38. Nakajima, N., M. Horikoshi, and R. G. Roeder. 1988. Factors involved in specific transcription by mammalian RNA polymerase II: purification, genetic specificity, and TATA box-promoter interactions of TFIID. Mol. Cell. Biol. 8:4028–4040.
39. Nunez, A.-M., M. Berry, J.-L. Imler, and P. Chambon. 1989. The 59 flanking region of the pS2 gene contains a complex enhancer region responsive to oestrogens, epidermal growth factor, a tumour promoter (TPA), the c-Haras oncoprotein and the c-jun protein. EMBO J. 8:823–829.

40. Ogryzko, V. V., R. L. Schiltz, V. Russanova, B. H. Howard, and Y. Nakatani. 1996. The transcriptional coactivators p300 and CBP are histone acetyltransferases. Cell 87:953–959.
41. Parekh, B. S., and T. Maniatis. 1999. Virus infection leads to localized hyperacetylation of histones H3 and H4 at the INF-B promoter. Mol. Cell 3:125–129.
42. Patikoglou, G. A., J. L. Kim, L. Sun, S.-H. Yang, T. Kodadek, and S. K. Burley. 1999. TATA element recognition by the TATA box-binding protein has been conserved throughout evolution. Genes Dev. 13:3217–3230.
43. Ruh, M. F., S. Tian, L. K. Cox, and T. S. Ruh. 1999. The effects of histone acetylation on estrogen responsiveness in MCF-7 cells. Endocrine 11:157–164.
44. Ryan, M. P., G. A. Stafford, L. Yu, and R. H. Morse. 2000. Artificially recruited TATA-binding protein fails to remodel chromatin and does not activate three promoters that require chromatin remodeling. Mol. Cell. Biol. 20:5847–5857.
45. Sabbah, M., K. Kang, L. Tora, and G. Redeuilh. 1998. Oestrogen receptor facilitates the formation of preinitiation complex assembly: involvement of the general transcription factor TFIIB. Biochem. J. 336:639–646.
46. Sadovsky, Y., P. Webb, G. Lopez, J. D. Baxter, P. M. Fitzpatrick, E. Gizang-
Ginsberg, V. Cavailles, M. G. Parker, and P. J. Kushner. 1995. Transcriptional activators differ in their responses to overexpression of TATA-box binding protein. Mol. Cell. Biol. 15:1554–1563.
47. Schiltz, R. L., C. A. Mizzen, A. Vassilev, R. G. Cook, C. D. Allis, and Y. Nakatani. 1999. Overlapping but distinct patterns of histone acetylation by the human co-activators p300 and PCAF within nucleosomal substrates. J. Biol. Chem. 274:1189–1192.
48. Sewack, G. F., and U. Hansen. 1997. Nucleosome positioning and transcription-associated chromatin alterations on the human estrogen-responsive pS2 promoter. J. Biol. Chem. 272:31118–31129.
49. Struhl, K. 1998. Histone acetylation and transcriptional regulatory mechanisms. Genes Dev. 12:599–606.
50. Torchia, J., C. Glass, and M. G. Rosenfeld. 1998. Co-activators and corepressors in the integration of transcriptional responses. Curr. Opin. Cell Biol. 10:373–383.
51. Tremblay, A., G. B. Tremblay, F. Labrie, and V. Giguere. 1999. Ligandindependent recruitment of SRC-1 to estrogen receptor b through phosphorylation of activator function AF-1. Mol. Cell 3:513–519.
52. Turner, B. M. 1993. Decoding the nucleosome. Cell 75:5–8.
53. Verdin, E., P. Paras, Jr., and C. V. Lint. 1993. Chromatin disruption in the promoter of human immunodeficiency virus type 1 during transcription activation. EMBO J. 12:3249–3259.

54. Verdone, L., G. Camilloni, E. D. Mauro, and M. Caserta. 1996. Chromatin-remodeling during Saccharomyces cerevisiae ADH2 gene activation. Mol. Cell. Biol. 16:1978–1988.
55. Widlund, H. R., J. M. Vitolo, C. Thiriet, and J. J. Hayes. 2000. DNA sequence-dependent contributions of core histone tails to nucleosome stability: differential effects of acetylation and proteolytic tail removal. Biochemistry 39:3835–3841.
56. Wolffe, A. P., and D. Pruss. 1996. Targeting chromatin disruption: transcription regulators that acetylate histones. Cell 84:817–819.
57. Workman, J. L., and R. E. Kingston. 1998. Alteration of nucleosome structure as a mechanism of transcriptional regulation. Annu. Rev. Biochem. 67:545–579.
58. Workman, J. L., and R. G. Roeder. 1987. Binding of transcription factor TFIID to the major late promoter during in vitro nucleosome assembly potentiates subsequent initiation by RNA polymerase II. Cell 51:613–622.
59. Workman, J. L., R. G. Roeder, and R. E. Kingston. 1990. An upstream transcription factor, USF (MLTF), facilitates the formation of preinitiation complexes during in vitro chromatin assembly. EMBO J. 9:1299–1308..
60. Wu, S.-Y., M. C. Thomas, S. Y. Hou, V. Likhite, and C.-M. Chiang. 1999. Isolation of mouse TFIID and functional characterization of TBP and TFIID in mediating estrogen receptor and chromatin transcription. J. Biol. Chem. 274:23480–23490.
61. Wu, C. 1997. Chromatin remodeling and the control of gene expression. J. Biol. Chem. 272:28171–28174.
62. Yokomori, K., M. P. Zeidler, J.-L. Chen, C. P. Verrijzer, and M. Mlodzik. 1994. Drosophila TFIIA directs cooperative DNA binding with TBP and mediates transcriptional activation. Genes Dev. 8:2313–2323.
63. Yoshida, M., M. Kijima, and T. Beppu. 1990. Potent and specific inhibition of mammalian histone deacetylase both in vivo and in vitro by trichostatin A. J. Biol. Chem. 265:17174–17179.

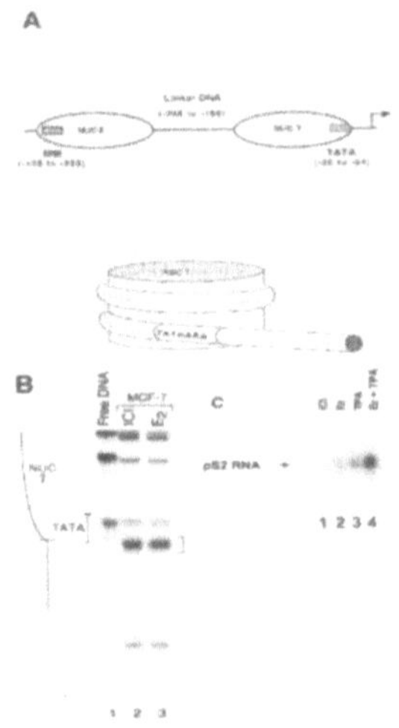

FIG. 1. The translational positioning of the nucleosome overlapping the TATA sequence is unchanged upon induction of the pS2 promoter. (A) The upper panel is a linear representation of the pS2 proximal promoter illustrating the positions of the estrogen response element (ERE) (hatched box) and TATA sequence (grey box) relative to the positions of the two nucleosomes (NUC E and NUC T). The transcriptional start site is denoted with an arrow. The lower panel further illustrates the positioning of the TATA sequence (nucleotides 230 to 224) of the human estrogen-responsive pS2 promoter within the rotationally and translationally positioned NUC T (nucleotides 223 to 2165) (48). (B) The boundary of NUC T is defined by chromatin-dependent micrococcal nuclease cleavages (bracketed on right) and is situated within the 39 edge of the TATA sequence (48). MCF-7 cells were treated with either the antiestrogen ICI 182,780 (ICI) or 17b-estradiol (E2) before chromatin was digested in vivo by addition of micrococcal nuclease. For comparison, purified genomic DNA (Free DNA) was digested in vitro and similarly analyzed. (C) Levels of pS2 RNA were determined by Northern blot analysis following treatment of MCF-7 cells for 48 h with ICI 182,780 (ICI) or for 3 h with 17bestradiol (E2) and/or TPA. Twenty micrograms of total isolated MCF-7 RNA per sample was electrophoresed through a 1.2% formaldehyde-agarose gel and transferred to a charged nylon membrane (NEN). The pS2 RNA was detected by hybridization

with a random primed DNA probe generated from the pS2 gene sequence.

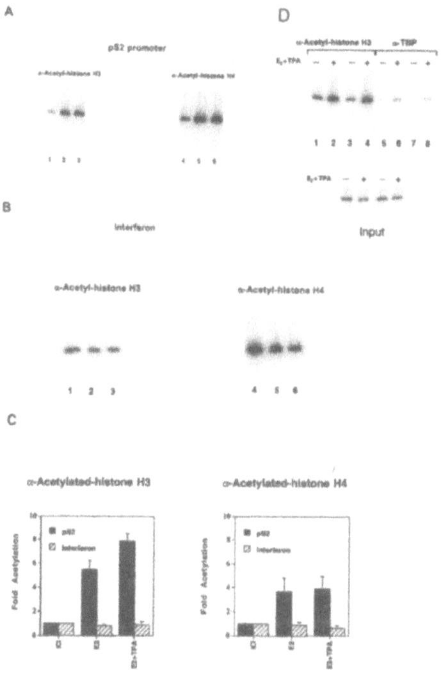

FIG. 2. Hyperacetylation of histones H3 and H4 and binding of TBP within the pS2 proximal promoter in response to the transcriptional stimuli estradiol and TPA. (A) Chromatin fragments were immunoprecipitated with anti-acetylated-histone H3 or H4 antibody as indicated, from human mammary MCF-7 cells treated with the pure antiestrogen ICI 182,780 (lanes 1 and 4), 17b-estradiol (lanes 2 and 5), or 17b-estradiol plus TPA (lanes 3 and 6). DNA purified from these immunoprecipitations was analyzed by quantitative PCR, with the product consisting of nucleotides 2440 to 118 of the pS2 promoter.(B) The same immunoprecipitates described for panel A (same lane designations) were analyzed by quantitative PCR with primer set F from the human beta interferon promoter region (2241 to 21) (41). (C) The relative levels of DNA immunoprecipitated with antibodies against acetylated histones H3 and H4 at the pS2 and beta interferon promoters were quantitated using a PhosphorImager (Molecular Dynamics). The average amounts of recovered DNA, with standard deviations,

were determined from three independent sets of experiments with separate batches of cells. Levels of DNA immunoprecipitated with anti-acetylated-histone H3 antibodies from ICI-treated cells were similar (within twofold) to background levels of precipitated DNA in the absence of antibodies. Thus, the actual enhancement in association of pS2 proximal promoter DNA with highly acetylated histone H3 may be even more pronounced. (D) Genomic chromatin fragments from MCF-7 cells, treated as described for panel A with either ICI 182,780 (lanes 1, 3, 5, and 7) or 17b-estradiol plus TPA (E2 1 TPA) (lanes 2, 4, 6, and 8), were immunoprecipitated either with anti-TBP antibodies (lanes 5 to 8) or with anti-acetylated-histone H3 antibodies (lanes 1 to 4). 2, absence of E2 1 TPA; 1, presence of E2 1 TPA. Duplicate experimental samples are shown to illustrate reproducibility. The relative levels of TBP binding to the proximal pS2 promoter and of immunoprecipitation with the anti-acetylated histone H3 antibody of this region were analyzed by quantitative PCR. The average increase in TBP binding with transcriptional stimuli was 4.1 6 1.1-fold over background levels, based upon four independent experiments. Quantitative PCR analysis was performed on DNA isolated from input samples prior to immunoprecipitation to confirm equalization of initial cellular material.

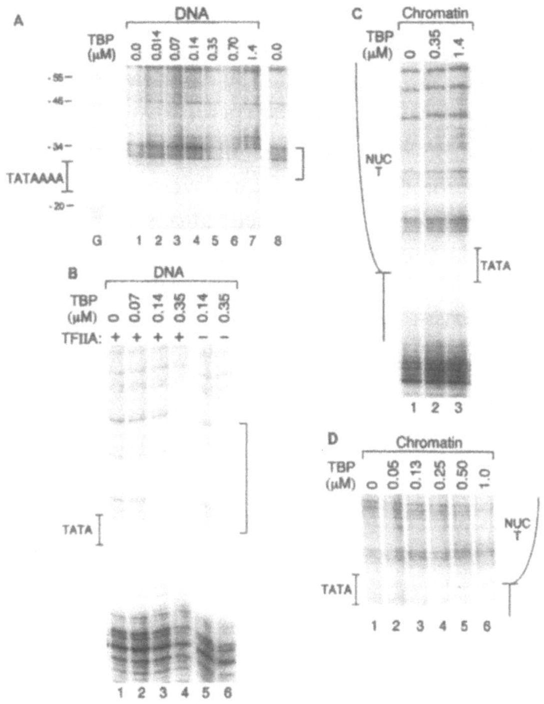

FIG. 3. The binding of TBP to the pS2 TATA sequence is severely restricted by the in vivo chromatin structure of the promoter. (A) Purified viral DNA containing the pS2 promoter (100 ng, 1.0 nM) was incubated with increasing concentrations of human TBP, at the concentrations (in micromolars) indicated above the individual lanes. Subsequently, the reaction mixtures were digested with DNase I, and the DNA products were amplified by LMPCR (48). TBP preparations were assayed for the percentages of TBP molecules that were active for specific DNA binding as described previously (25). The preparation used for the experiments illustrated in Fig. 3D, 4A, and 5A contained essentially all active protein, while the preparation used for the experiments shown in Fig. 3A to C was approximately 30% as active. A control DNA sample was incubated in the absence of TBP (lanes 1 and 8). Dimethyl sulfate/piperidine-treated SVSpS2 viral DNA provided a G ladder (lane G); the nucleotide positions of the pS2 promoter are indicated to the left of this lane. The position of

the TATA box is illustrated by the pS2 sequence TATAAAA. The binding of TBP is evidenced by a region of protection from cleavage as delineated by the bracket to the right. (B) Purified viral DNA containing the pS2 promoter was incubated with TBP at the protein concentrations indicated above the individual lanes but supplemented either with the basal transcription factor TFIIA (1) or with TFIIB (lanes 5 and 6). TFIIA and TFIIB concentrations equaled or exceeded that of TBP; TFIIA concentrations were as follows: lanes 1 and 2, 0.13 mM; lane 3, 0.25 mM; lane 4, 0.63 mM. TFIIB concentrations were as follows: lane 5, 0.14 mM; lane 6, 0.35 mM. The designations are as given for panel A. (C) MCs (230 pM) containing the pS2 promoter were isolated from CMT cells and were incubated with TBP at the concentrations noted. The position of NUC T is diagrammed on the left. The remaining designations are as noted above. (D) MCs were incubated with TBP as described for panel C, with the addition of the basal transcription factor TFIIA. TFIIA concentrations were as follows: lanes 1 to 4, 0.25 mM; lane 5, 0.50 mM; lane 6, 1.0 mM.

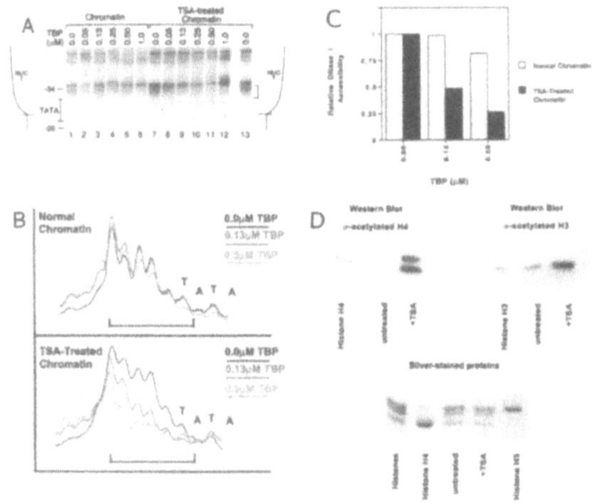

FIG. 4. MCs isolated from TSA-treated cells support enhanced levels of TBP binding to the pS2 TATA sequences within NUC T. (A) MCs (230 pM) isolated from either untreated cells (lanes 1 to 6) or TSA-treated cells (lanes 7 to 13) were incubated with increasing concentrations of TBP for DNase I footprinting. These same MC preparations were analyzed for panel D for the acetylation state of histones H3 and H4. The concentration of TBP is indicated above each

lane. TFIIA concentrations were as follows: lanes 1 to 4, 7 to 10, and 13, 0.25 mM; lanes 5 and 11, 0.50 mM; lanes 6 and 12, 1.0 mM. The position of NUC T is diagrammed on both sides of the lanes. The nucleotide positions within the pS2 promoter are numbered to the left. The region of highly specific, TBP-dependent protection from DNase I cleavage (bracket) is indicated to the right. The results shown in this figure are representative of four independent experiments. (B) The bands within the TBP-dependent footprint region (panel A, bracket) were scanned using a PhosphorImager (Molecular Dynamics). The upper graph illustrates the TBP interaction with normal chromatin (panel A, lanes 1, 3, and 5); the lower graph illustrates the TBP interaction with TSA-treated chromatin (Fig. 4A, lanes 7, 9, and 11). The positions of the nucleotides 230 to 227 (TATA) are indicated above their respective peaks. The x axis represents the relative position within the gel, while the y axis illustrates band intensity. The brackets denote the region of DNase I protection. (C) The relative DNase I accessibility of the TBP-dependent footprint region was compared for the two preparations of chromatin. The intensities of DNase I cleavage within the TBP footprint, denoted by brackets in panels A and B, were quantitated and corrected for total lane intensity by dividing this measurement by the intensity of a region within the lane unaffected by the binding of TBP (3). The accessibility without the addition of TBP was arbitrarily set at 1, with increased protection by TBP resulting in lower accessibility to DNase I. DNA sequences that became hypersensitive to DNase I were not included as part of the footprinted region. (D) Western blot analysis of the acetylation level of histones H4 and H3, from MCs isolated from either untreated or TSA-treated (1TSA) cells (preparations used for panel A). The indicated antibodies were used to probe the histone modification state. The anti-acetylated-histone H4 rabbit antiserum also interacts with acetylated histone H2B under conditions of substantial excess antibody (Upstate Biotechnology Co., personal communication). Marker lanes contained high-performance liquid chromatography-purified chicken erythrocyte histones H3 and H4. MC samples were also analyzed by staining with silver (bottom panel). The amounts of histones in the MCs from untreated versus TSA-treated cells were comparable. However, TSA treatment led to an accumulation of hyperacetylated histones H3 and H4.

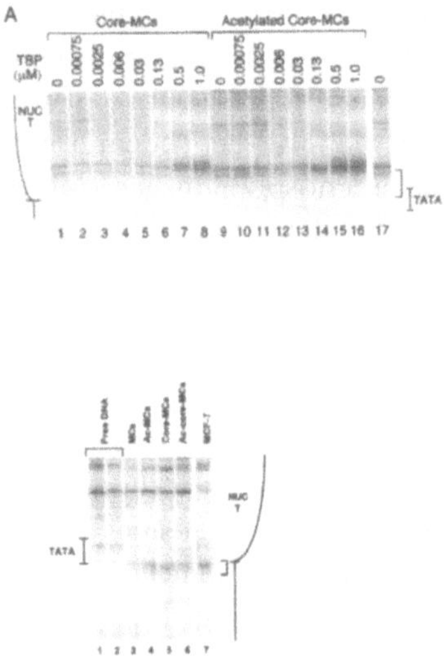

FIG. 5. Binding of TBP to NUC T is facilitated by acetylation of core histones. (A) Core MCs, containing only the core histones assembled in vivo with recombinant pS2 promoter-containing viral DNA, were prepared from MCs isolated from either untreated or TSA-treated, infected cells. The core MC preparations were incubated prior to DNase I digestion with increasing concentrations of TBP as shown. TFIIA concentrations are as follows: lanes 1 to 6, 9 to 14, and 17, 0.25 mM; lanes 7 and 15, 0.50 mM; lanes 8 and 16, 1.0 mM. Control reaction mixtures were incubated without TBP (lanes 1, 9, and 17). The designations are as given for Fig. 4A. The results shown in this figure are representative of four independent experiments. The detailed pattern of DNase I cleavages near the TATA sequence obtained here differed slightly from the pattern obtained for Fig. 3C and D and 4A. In our experience, the final step of the LMPCR procedure that generates the radiolabeled DNA products yields slight variability from experiment to experiment in the pattern of relative intensities of different bands. However, within any single experiment, consistency in the detailed DNase I cleavage patterns is observed. (B) The translational positioning of NUC T is strictly maintained on the salt-treated core MCs. MCs and core MCs, both normal and more highly acetylated (Ac) (lanes 3 to 6), were digested with micrococcal nuclease (0.05 U) prior to LMPCR amplification to establish

the boundary of NUC T. In vivo micrococcal nuclease cleavage of genomic DNA within MCF-7 human breast cells, followed by LMPCR (48), demonstrates the in vivo translational positioning of NUC T (lane 7). Micrococcal nuclease cleavage (lane 1, 0.005 U; lane 2, 0.01 U) of purified recombinant viral DNA (Free DNA) is shown as controls. The enhanced cleavages present in all the chromatin preparations but not in the free DNA are delineated with a bracket. The positions of NUC T and the TATA box are indicated schematically at the sides of the lanes.

Appendix C

Article No. mc980738
Review

Chromatin Structure and Cardiac Gene Expression

Weei-Yuarn Huang and Choong-Chin Liew
Laboratory for Molecular Cardiology, Departments of Laboratory Medicine and Pathobiology, University of Toronto, Toronto, Ontario, Canada
(Received 21 January 1998, accepted in revised form 14 May 1998)

W.-Y. HUANG, AND C.-C. LIEW. Chromatin Structure and Cardiac Gene Expression. *Journal of Molecular and Cellular Cardiology* (1998) **30**, 1673–1681.

Gene activation is often preceded by or accompanied by a perturbation of the chromatin structure. Recently, several co-factors of transcription factors have been identified as histone acetyltransferases. In addition, retinoblastoma protein and some co-factors can form a complex with histone deacetylases. These discoveries provide direct evidence that modification of chromatin structure is crucial for gene activation. However, the role of chromatin structure in cardiac-specific expression has not yet been elucidated. The potential significance of chromatin structure in cardiac-specific gene expression is indicated by: (1) heterogeneous human SWI/SNF chromatin remodelling factors among various tissues; (2) several chromatin remodelling factors shown to be expressed preferentially in the heart; (3) the demonstration of chromatin remodelling of the cardiac b-myosin heavy chain gene (MyHC) during cardiac development. We therefore propose to study cardiac-specific chromatin remodelling activity in order to elucidate mechanisms controlling the reactivation of the fetal heart genetic program in the hypertrophic heart. © 1998 Academic Press

KEY WORDS: Chromatin remodelling factors; Transcription factors; DNase I hypersensitive sites; Cardiac hypertrophy.

Introduction

Gene expression in eukaryotes is subject to multiple levels of control. Transcription initiation hasbeen the focus of intense research, and much progress has been made *in vitro* in understanding this aspect of gene regulation (for review see Orphanides *et al.*, 1996). Unlike the naked DNA used in those *in vitro* transcription experiments, eukaryotic DNA is not freely accessible *in vivo*, but is organized with histone and non-histone proteins to form a chromatin fiber. It was previously thought that the primary role of chromatin was to act as a passive repressor in transcription. However, recent biochemical and genetic evidence indicates that chromatin is a dynamic structure. Upon gene activation, chromatin structure is changed through the rearrangement of histones and DNA. Eukaryotic gene expression can be understood only in the context of these changes (Felsenfeld, 1972; Lewin, 1994; Svaren and Horz, 1996; Wolffe and Pruss, 1996).

The fundamental packing unit of chromatin, nucleosome, gives chromatin a beads-on-a-string appearance in electron micrographs taken after treatments that unfold higher-order packing. Two copies of each histone protein, H2A, H2B, H3 and H4, are assembled into an octamer that has 146 base pair (bp) of DNA wrapped around it to form a nucleosome core (Luger *et al.*, 1997). Regulatory elements of several genes were found to be organized with specifically positioned nucleosomes (Simpson, 1986; Travers, 1987). Such a unique structure has been indicated as an intrinsic part of the regulatory mechanism for eukaryotic genes. The mouse mammary tumor virus (MMTV) (Richard-Foy and Hager, 1987; Truss *et al.*, 1995) and the yeast PHO 5 gene (Almer and Horz, 1986; Schmid *et al.*, 1992) are two well-characterized models showing that their promoters are packaged nucleosomes. Gene activation is often accompanied by perturbations of the arrays of nucleosomes that make regions of promoters and enhancers more sensitive to nuclease digestion (i.e. the formation of nuclease hypersensitive site) (Wallrath *et al.*, 1994).

Nuclease hypersensitive site

In chromatin, nuclease hypersensitive sites arebelieved to represent "open windows" that allow enhanced access of crucial *cis*-acting DNA sequences to *trans*-acting factors. These accessible regions are defined by their pronounced sensitivity to nuclease cleavage (such as DNase I, restriction enzymes, micrococcal nuclease, or endogenous nuclease) or chemical modification, and are typically two orders of magnitude more sensitive than are other regions in bulk chromatin (for review see Gross and Garrard, 1988). A wide variety of functional sequences (e.g. enhancer, silencer, or promoter) are associated with hypersensitive sites (tabulated see Gross and Garrard, 1988).

Showing that active sites are DNase I-hypersensitive does not tell us much about the nature of these sites. Indeed, the exact mechanisms for hypersensitivity and its role in gene regulation are only partially understood at present. Based on low-resolution nuclease-digestion experiments, it was concluded that these sites represent nucleosome-free regions of DNA (Almer and Horz, 1986; Elgin, 1988; Becker, 1994). However, studies of the MMTV promoter and the albumin gene's enhancer led to a different discovery, i.e. the persistence of nucleosomes within the DHS (McPherson *et al.*, 1993; Truss *et al.*, 1995). Using high-resolution *in vivo* footprinting analysis, Mcpherson *et al.* (1993) demonstrated that an albumin distal enhancer exists in an array of three precisely positioned nucleosomes only in liver chromatin, where the enhancer is active. This enhancer is hypersensitive exactly between the micrococcal nuclease cleavages that define one of the nucleosomal particles. Also, Truss *et al.* (1995) demonstrated that hormone induction in MMTV promoter causes a DHS limited to a short region in the centre of nucleosome B; this nucleosome is neither removed nor shifted upon hormone induction. Hence, it appears that either possibility (free or persistent of nucleosome) may exist at different DHS in cellular chromatin.

Architectural variations of inducible eukaryotic promoters: preset and remodelling chromatic structures

Inducible genes have two types of chromatin stucture: preset and remodelling (Wallrath *et al.*, 1994). Preset genes are those in which the binding sites for *trans*-acting factors are accessible prior to activation (i.e. in a DNase I hypersensitive configuration). In response to the activation signal, *trans*-acting factors bind to *cis*-acting regulatory elements and trigger transcription with no major alterations in the chromatin structure of the promoter region. By contrast, remodelling genes are those in which some of the required *cis*-acting regulatory elements are packaged into nucleosomes (i.e. in a non-DNase hypersensitive configuration). In order for the *trans*-acting factors to gain access to *cis*-acting elements, the nucleosomes must be perturbed in response to an activation signal; a chromatin remodelling process, which forms DNase I, hypersensitive sites must occur. Furthermore, it has been demonstrated that the alteration of nucleosomal structure observed in both systems is not due to the consequences of ongoing transcription (Fascher *et al.*, 1993; Truss *et al.*, 1995). Thus, perturbation of nucleosomal structure appears to be a prerequisite for, or an initial step in, gene activation.

Mechanisms of chromatin remodelling

(1) Transcription factors bind to nucleosomes

An important event in the remodelling process appears to be the initial binding of a positive regulator to its site over a nucleosome (Wallrath *et al.*, 1994). However, not all transcription factors can bind to their sites on the face of a nucleosome. While the glucocorticoid receptor (GR) can bind to a nucleosomal site with almost the same affinity as it binds to free DNA, the NF1 and Oct 1 transcription factors cannot bind to the positioned nucleosome over MMTV promotor before hormone induction (Archer *et al.*, 1992). It is unclear whether this difference in binding ability is due to the locations of the binding sites of nucleosomes or to distinct properties of the proteins. However, the GAL4 and HSF appear to be mechanisticially distinct (Taylor *et al.*, 1991), that is, the HSF can bind only to free DNA, whereas, the GAL4 can bind to randomly positioned or rotationally phased nucleosomes *in vitro*.

(2) Modification of histones (Fig. 1)

Histones may be modified in two ways: by alterations in histone composition (such as depletion of linker histone H1 and histone H2A/H2B dimers) (Hayes and Wolffe, 1992), or by chemical modifications (e.g. histone hyperacetylation). Acetylation occurs at lysine residues on the amino-terminal tails of the histones thereby reducing their net positive charge and decreasing their affinity for DNA. It has been recognized for many years that core histone hyperacetylation is correlated to gene activation, and the elimination of histone acetylation is correlated with transcriptional silencing (for review see Struhl, 1998). However, there was essentially no information about the underlying molecular mechanisms of histone acetylation and transcriptional activity.

The first mechanistic connection between histone acetylation and transcriptional regulation came from the identification that several co-activators are hist-sequence- one acetyltransferases (Bannister and Kouzarides, 1996; Brownell *et al.*, 1996; Mizzen *et al.*, 1996; Ogryzko *et al.*, 1996; Yang *et al.*, 1996). Three mammalian coactivators, p300/CBP, P/CAF and TAFII 250, have been identified as histone acetylremains transferases, indicating that sequence-specific activators or TATA-binding protein is able to overcome the repressive effect of chromatin by directly recruiting one of its cofactors to modify nucleosomal structure.

Transcription factors can also modify nucleosomal structure via the recruitment of histone deacetylase (HDAC). So far, three histone deacetylases have been identified (HDAC1, 2 and 3) (Yang *et al.*, 1997). HDAC1 was found to be associated with N-CoR (nuclear receptor corepressor) and Sin3, which is mamable to repress transcription *in vivo* (Heinzel *et al.*, 1997). Since Nco-R and p300 is a corepressor and a coactivator of thyroid hormone receptor (TR), respectively, it is proposed that nuclear receptor can reversibly modify the chromatin structure of its target genes through the recruitment of histone acetyltransferase or deacetylase to the promoter regions (Wolffe, 1997). Consistent with this model, the tumor suppressor gene, retinoblastoma protein, was also shown to repress gene expression via the recruitment of histone deacetylase (Brehm *et al.*, 1998; Luo *et al.*, 1998; Magnaghi-Jaulin *et al.*, 1998).

(3) Recruitment of SWI-SNF complex (or other nucleosome remodelling factors)

Several chromatin remodelling factors have been identified, including SWI-SNF complex (SWI for (nucleosome remodelling factor) (Tsukiyama and Wu, 1995), and RSC (Remodel the Structure of Chromatin) (Cairns *et al.*, 1996a). The yeast SWI-SNF complex is a conserved multi-subunit complex consisting of 11 proteins (for review see Winston and Carlson, 1992; Peterson and Tamkun, 1995). One of the subunits of the SWI-SNF complex, SW12–SNF2, has been characterized as a DNA-stimulated ATPase and may be the motor of the nucleosome remodelling machinery. Interestingly, NURF and RSC also contain a SW12–SNF2-like subunit.

Wilson *et al.* (1996) reported an intriguing finding that SWI–SNF proteins are also integral components of the *S. cerevisiae's* RNA polymerase II holoenzyme. Since the RNA polymerase II holoenzyme is required for many yeast genes *in vivo* (Thomson and Young, 1995), it was proposed that sequence-specific activators are able to recruit the holoenzyme, and hence the SWI–SNF components to the promoter region (Peterson, 1996). Although whether the SWI–SNF complex is associated with RNA polymerase II holoenzyme remain-san issue of controversy (Cairns *et al.*, 1996a), this model is attractive in that it provides a simple solution to the question of how SWI–SNF proteins are brought to promoters (Peterson, 1996).

The purification and cloning of mammalian SWI–SNF complexes was achieved by Wang *et al.* (1996a,b). The subunits that they cloned are similar to the yeast counterparts. While the observed similarity between mammalian and yeast complexes is significant, the complexes in mammals are more diverse than the SWI–SNF complex in yeast. The BAF60 subunit is a member of a novel gene family. Its yeast homology, SWP73, is required for GR activation in yeast (Cairns *et al.*, 1996b). Three members of this family with different tissue-specific expression patterns were cloned, BAF60a, BAF60b, and BAF60c. While BAF60a is expressed at similar level in various tissues examined, the BAFb and BAFc are predominantly expressed in pancreas and muscle (including heart and skeletal muscle), respectively. Furthermore, two other subunits, BAF155 and BAF170, are also expressed predominantly in the cardiac and skeletal muscles. The relatively selective expression of some subunits in different tissues suggests the existence of tissue-specific chromatin remodelling activity in mammals.

Chromatin remodelling of the cardiac myosin heavy chain genes

Several cardiac genes have been analysed for DNase I hypersensitive sites. They are chicken light chain gene (MLC2A) (Winter and Arnold, 1987), cardiac actin gene (Biben *et al.*, 1994), and cardiac a- and b- myosin heavy chain genes (MyHC) (Huang and Liew, 1997; Huang *et al.*, 1997). Using nuclei isolated from chicken embryonic heart, Winter and Arnold (1987) first identified four cardiac-specific DHSs in the MLC2A gene, and concluded that these sites may be related to the expression of this gene. Biben *et al.* (1994) using nuclei isolated from skeletal muscle cells, identified two DHSs in the 5' flanking region of the mouse a-cardiac actin gene. Increasing intensity of these DHSs was observed during myoblastic differentiation into myotubes (Biben *et al.*, 1994). Although these studies demonstrated the existence of DHSs in the cardiac-specific genes, they did not address whether, during mammalian cardiogenesis, there is a change in chromatin structure accompanying the cardiacp300/specific gene expression.

In order to explore the role of chromatin remodelling in cardiac gene expression, we studied the DHS pattern of the hamster cardiac MyHC genes during cardiac development. The mammalian cardiac α- and β-MyHC genes are tandemly aligned in the genome (i.e. β-α) and separated by a 4-kb intergenic region (Liew *et al.*, 1990; Epp *et al.*, 1993). The cardiac MyHC genes are expressed in a tissue- and developmentally-specific manner, and are primarily regulated at the transcriptional level (Lompre *et al.*, 1984; Boheler *et al.*, 1992). In the cardiac ventricle of rodents, the b-MyHC transcripts predominate during embryonic and fetal development. At about the time of birth, a circulating thyroid hormone surge causes a switch in cardiac MyHC gene expression, characterized by the down regulation of β-MyHC gene expression, and a concomitant induction of the α-isoform. Pathological processes also influence the regulation of these genes. For example, cardiac hypertrophy causes the re-expression of the β-MyHC gene in the ventricles of rodents (Izumo *et al.*, 1987; Waspe *et al.*, 1990).

The proximal regulatory elements of both cardiac α- and β-MyHC genes have been identified by transient transfection and footprinting analysis (Mably and Liew, 1996). For the l-MyHC gene, these include the thyroid hormone (T3) response elements (TRE) (Flink and Morkin, 1990), the DNA-binding sites of MEF-2 (Molkentin and Markham, 1993), myogenin factors (Molkentin *et al.*, 1993), M-CAT binding factor (TEF-1) (Molkentin and Markham, 1994), and GATA-4 (Molkentin *et al.*, 1994). The regulatory elements identified in the cardiac β-MyHC gene promoters of several mammalian species are summarized in Figure 2. Unlike the well-defined TREs of the α-MyHC gene, the TRE responsible for inhibiting β-MyHC expression remains undefined by transient transfection analysis (Morkin,1993). In addition, the functional significance of these regulatory elements has been examined by transgenic studies. Although those data are consistent with *in vitro* cell culture studies, they further suggested that far upstream regulatory elements play an important role in the regulation of the cardiac MyHC gene expression (Subramaniam *et al.*, 1993; Knotts *et al.*, 1994).

Consistent with transgenic studies, we identified a cardiac-specific DHS located in the 4-kb upstream region of the cardiac α-MyHC gene (Huang and Liew, 1997). When using nuclei isolated from late-gestational and adult heart ventricles, this DHS was inmapped to the -1.9-kb region. Within this site, we found a conserved GATA-motif that interacts specifically with GATA-binding factors in nuclear extracts of cardiomyocytes at various developmental stages. These data provide further evidence to support the role of GATA factors in the regulation of cardiac α-MyHC gene expression.

Two cardiac-specific DHSs were identified within the 5-kb upstream region of the cardiac β-MyHC gene. One of the DHSs was mapped to -2.3-kb (β-2.3-kb) region and the other to the proximal promoter region (Huang *et al.*, 1997). Although these two sites were readily detectable using nuclei from neonatal hamster heart, the proximal promoter site disappeared when adult hamster heart nuclei were used, and the β-2.3-kb site decreased in intensity [Fig. 3(a)]. This alteration in the DHS pattern corresponds to the shift in expression of the cardiac MyHC genes in the hamster (i.e. from β to α). Furthermore, we were unable to detect the promoter DHS in the other tissues [Fig. 3(a)] or non-cardiomyocytes (Huang *et al.*, 1997), suggesting that the existence of this DHS in the heart GATA could be a result of tissue-specific chromatin remodelling activity. Figure 3(b) summarizes our results of both DNase I and micrococcal nuclease digestion of the cardiac β-MyHC gene within intact heart nuclei (Huang and Liew, 1998).

Cardiac transcription factors and chromatin remodelling

The cardiac transcription factors play a pivotal role in cardiogenesis and cardiac-specific gene regulation (for review see Mably and Liew, 1996). By using gene-targeting analysis, cardiac transcription factors, GATA-4 (Kuo *et al.*, 1997; Molkentin *et al.*, 1997), MEF2C (Lin *et al.*, 1997), Csx/Nkx-2.5 (Lyons *et al.*, 1995), TEF-1 (Chen *et al.*, 1994), dHAND (Srivastava *et al.*, 1997) and eHAND (Firulli *et al.*, 1998) have been shown to be essential for cardiac development. Several lines of evidence indicate that these essential transcription factors are also involved in cardiac hypertrophy. For example, GATA-4 was shown to mediate the reactivation of the β-MyHC gene in a cardiac hypertrophy model (Hasegawa *et al.*, 1997). More significantly, GATA-4 was recently shown to link to NF-AT3, a downstream transcription factor of calcineurin signaling pathway (Hoey *et al.*, 1995). Cardiac-specific expression of calcineurin in transgenic mice causes cardiac hypertrophy (Molkentin *et al.*, 1998).

Based on the chromatin remodelling observed in the cardiac β-MyHC gene, it is hypothesized that the fetal cardiac genes may adopt a "closed" chromatin structure in the adult heart stage. Hence, even though the cardiac nuclear factors are highly expressed, they are unable to get access into their target genes. In order to activate fetal gene expression, cardiac transcription factors first have to overcome the chromosomal barrier. At present, we do not know whether GATA-4 is capable of binding to nucleosomal DNA or not. However, GATA-

1 (which contains a highly homologous DNA-binding domain with GATA-4), is unable to bind to its binding site over the chromatin-reconstituted β-globin gene *in vitro* (Armstrong and Emerson, 1996). Only after this nucleosomal DNA was disrupted by binding to another erythroid-specific nuclear factor (NF-E2), was GATA-1 allowed to bind to its binding site. Thus, it is possible that synergistic activation of the ANF (atrial natriuretic factor) promoter by GATA-4 and NF-AT3 (observed by Molkenten *et al.*, 1998) results from a similar mechanism in which NF-AT3 is recruited to facilitate GATA-4 to bind to its site over nucleosomal DNA. However, this model remains to be tested.

Future Directions

Whether chromatin remodelling is applied to other fetal cardiac genes remains to be determined. It might be interesting to investigate the chromatin structure of the ANF gene, since the developmental expression pattern of this gene is somewhat similar to that of the cardiac β-MyHC gene. The ANF mRNA is expressed throughout embryonic and fetal development in atrial and ventricular cells (Zeller *et al.*, 1987). However, in the latter stages of fetal development, the ANF gene is switched off in ventricular cells, whereas its level of expression remains high in atria. The decrease in ventricular ANF expression occurs around the time of birth in mammals, and it is re-expressed in hypertrophied ventricular cells (Chien *et al.*, 1991). Also, a functional GATA motif was identified in the ANF proximal promoter region (Grepin *et al.*, 1994).

The mammalian SWI–SNF complex is heterogeneous, with some subunits preferentially expressed in heart (Wang *et al.*, 1996a,b). Four histone acetyltransferases (HAT) have been identified and more HATs await discovery (Yang *et al.*, 1996; Ogryzko *et al.*, 1996). In one of them, P/CAF mRNA is most abundant in cardiac and skeletal muscle (Yang *et al.*, 1996). Therefore, it is speculated that the heart may adopt a unique chromatin remodelling mechanism in gene regulation.

Since activation of fetal cardiac genes is the target of the hypertrophic signals, cardiac-specific chromatin remodelling factors should act as essential factors in response to hypertrophic signals. Thus, the identification of the cardiac-specific chromatin remodelling factors and determination of which cardiac nuclear factors are capable of recruiting this chromatin remodelling machinery will be crucial to our understanding of the reactivation of fetal genetic program in cardiac hypertrophy. This cardiac-specific chromatin remodelling activity may be an alternative therapeutic target for preventing (or slowing down) the progression of cardiac hypertrophy.

Acknowledgments
This work was supported by the Ontario Heart and Stroke Foundation and the Medical Research Council of Canada. Weei-Yuarn Huang was supported by a University of Toronto Open Fellowship.

References

Almer A, Horz W, 1986. Nuclease hypersensitive regions with adjacent positioned nucleosomes mark the gene boundaries of the PHO5/PHO3 locus in yeast. *EMBO J* **5:** 2681–2687.

Archer TK, Lefebvre P, Wolford RG, Hager GL, 1992. Transcription factor loading on the MMTV promoter: A bimodal mechanism for promoter activation. *Science* **255:** 1573–1576.

Armstrong JA, Emerson BM, 1996. NF-E2 disrupts chromatin structure at human β-globin locus control region hypersensitive site 2 *in vitro*. *Mol Cell Biol* **16:** 5634–5644.

Bannister A, Kouzarides T, 1996. The CBP co-activator is a histone acetyltransferase. *Nature* **384:** 641–643.

Becker PB, 1994. The establishment of active promoters in chromatin. BioEssays **16:** 541–547.

Biben C, Kirschbaum BJ, GarnerI, Buckingham M, 1994. Novel muscle-specific enhance sequences upstream of the cardiac actin gene. *Mol Cell Biol* **14:** 3504–3513.

Boheler KR, Chassagne C, Martin X, Wisnewsky C, Schwartz K, 1992. Cardiac expression of α- and β-myosin heavy chains and sarcomeric α-actins are regulated through transcriptional mechanisms. *J Biol Chem* **267:** 12979–12985.

Brehm A, Miska EA, McCance DJ, Reid JL, Bannister AJ, Kouzarides T, 1998. Retinoblastoma protein recruits histone deacetylase to repress transcription. *Nature* **391:** 597–601.

Brownell JE, Zhou J, Ranalli T, Kobayashi R, Edmondson DG, Roth SY, Allis CD, 1996 Tetrahymena histone acetyltransferase A: a homolog to yeast Gcn5p linking histone acetylation to gene activation. *Cell* **83:** 843–851.

Cairns BR, Lorch Y, LI Y, Zhang M, Lacomis L, Erdjument-Bromage H, Tempst P, DU J, Laurent B, Korenberg RD, 1996a. RSC, an essential, abundant chromatin-remodelling complex. *Cell* **87:** 1249–1260.

Cairns BR, Levinson RS, Yamamoto KR, Kornberg RD, 1996b. Essential role of Swp73p in the function of yeast Swi/Snf complex. *Gene Dev* **10:** 2131–2144.
Chen Z, Friedrich GA, Soriano P, 1994. Transcriptional enhancer factor 1 disruption by a retroviral gene trap leads to heart defects and embryonic lethality in mice. *Genes Dev* **8:** 2293–2301.
Chien KR, Knowlton KU, Zhu H, Chien S, 1991. Regulation of cardiac gene expression during myocardial growth and hypertrophy: molecular studies of an adaptive physiologic response. *FASEB J.* **5:** 3037–3046.
Elgin SCR, 1988. The formation and function of DNase I hypersensitive sites in the process of gene activation. *J Biol Chem* **263:** 19259–19262.
Epp TA, Dixon IC, Wang H-Y, Sole MJ, Liew C-C, 1993. Structural organization of the human cardiac α-MyHC gene. *Genomic* **18:** 505–509.
Fascher KD, Schmitz J, Horz W, 1990. Role of trans-acting proteins in the generation of active chromatin at the PHO5 promoter in *S. Cerevisiae. EMBO J.* **9:** 2523–2528.
Felsenfeld G, 1992. Chromatin as an essential part of the transcriptional mechanism. *Nature* **355:** 219–224.
Firulli AB, McFadden DG, Lin Q, Srivastava D, Olson E, 1998. Heart and extra-embryonic mesodermal defects in mouse embryos lacking the bHLH transcription factor. *Nat Genet* **18:** 266–270.
Flink IL,Morkin E, 1990. Interaction of thyroid hormone receptors with strong and weak *cis*-acting elements in the human α-myosin heavy chain promoter. *J Biol Chem* **265:** 11233–11237.
Flink IL, Morkin E, 1995. Alternatively processed isoforms of cellular nucleic acid-binding protein interact with a suppressor region of the human b-myosin heavy chain gene. *J Biol Chem* **270:** 6959–6965.
Flink IL, Edwards JG, Bahl JJ, Liew C-C, Sole MJ, Morkin E, 1992. Characterization of a strong positive *cis*-acting element of the β-myosin heavy chain gene in fetal rat heart cells. *J Biol Chem* **267:** 9917–9924.
Grepin C, Dagnino L, Robitaille L, Haberstroh L, Antakly T, Nemer M, 1994. A hormone-encoding gene identifies a pathways for cardiac but not skeletal muscle gene transcription. *Mol Cell Biol* **14:** 3115–3129.
Gross DS, Garrard WT, 1988. Nuclease hypersensitive sites in chromatin. *Annu Rev Biochem* **57:** 159–197.
HasegawaK, Lee SJ, Jobe SM, Markham BE, Kitsis RN, 1997. *Cis*-acting sequcnces that mediate induction of the β-myosin heavy chain gene expression

during left ventricular hypertrophy due to aortic constriction. *Circulation* **96:** 3943–3953.

Hayes JJ, Wolffe AP, 1992. Histones H2A/H2B inhibit the interactions of transcription factor IIIA with the *Xenopus borealis somatic* 5S RNA in a nucleosome. *Proc Natl Acad Sci* **89:** 1229-1233.

Heinzel T, Lavinsky RM, Mullen T-M, Soderstrom M, Laherty CD, TorchiaJ, Yang W-M, Brard G, Ngo SD, Davie JR, Seto E, Eisenman RN, Rose DW, Glass CK, Rosenfeld MG, 1997. A complex containing N-CoR, mSin3 and histone deacetylase mediates transcriptional repression. *Nature* **387:** 43–48.

Hoey T, Sun Y-L, Williamson K, Xu X, 1995. Isolation of two new members of the NF-AT gene family and functional characterization of the NF-AT proteins. Immunity **2:** 461–472.

Huang W-Y, Liew C-C, 1997. A conserved GATA motif in a tissue-specific DNase I hypersenstive site of the cardiac α-myosin heavy chain gene. *Biochem J* **325:**47–51.

Huang W-Y, Liew C-C, 1998. Chromatin remodelling of the cardiac β-MyHC gene. *Biochem J* **330:** 871–876.

Huang W-Y, Chen J-J, Shih N-L, Liew CC, 1997. Multiple muscle-specific regulatory elements are associated with a DNase I hypersensitive sites of the cardiac β-myosin heavy chain gene. *Biochem J* **327:** 507–512.

Izumo S, Lompre A-M, Matsuoka R, Koren G, Schwartz K, Nadal-Ginard B, Mahdavi V, 1987. Myosin heavy chain messenger RNA and protein isoform transitions during cardiac hypertrophy. *J Clin Invest* **79:** 970–977.

Kariya K, Farrance IKG, Simpson PC, 1993. Transcriptional enhancer factor-1 in cardiac myocytes interacts with an alpha 1 adrenergic- and a beta-protein kinase C-inducible element in the rat beta-myosin heavy chain promoter. *J Biol Chem* **268:** 26658–26662.

Knotts S, Rindt H, Neumann J, Robbins J, 1994. *In vivo* regulation of the mouse β-myosin heavy chain gene. *J Biol Chem* **269:** 31275–31282.

Kuo CT, Morrisey EE, Anandappa R, Sigrist K, Lu MM, Parmacek MS, Soudais C, Leiden JM, 1997. GATA4 transcription factor is required for ventral morphogenesis and heart tube formation. *Genes Dev* **11:** 1048–1060.

Lewin B, 1994. Chromatin and gene expression: constant questions, but changing answers. *Cell* **79:** 397–406.

Liew C-C, Sole M-J, Yamauchi-Takihara K, Kellam B, Anderson DH, LIN L, LIEW JC, 1990. Complete sequence and organization of the human cardiac b-myosin heavy chain gene. *Nucleic Acids Res* **18:** 3647–3651.

Lin Q, Schwarz J, Bucana C, Olson EN, 1997. Control of mouse cardiac morphogenesis and myogenesis by transcription factor MEF2C. *Science* **276:** 1404–1407.

Lompre AM, Nadal-Ginard B, Mahdavi V, 1984. Expression of the cardiac ventricular alpha and beta-myosin heavy chain genes is developmentally and hormonally regulated. *J Biol Chem* **259:** 6437–6446.

Luger K, Mader AW, Richmond RK, Sargent DF, Richmond TJ, 1997. Crystal structure of the nucleosome core particle at 2.8 A resolution. *Nature* **389:** 251–260.

Luo RX, Postigo AA, Dean D, 1998. RB interacts with histone deacetylase to repress transcription. *Cell* **92:** 463–473.

Lyons I, Parsons LM, Hartley L, LI R, Andrews JE, Robb L, Harvey RP, 1995. Myogenic and morphogenetic defects in the heart tubes of murine embryos lacking the homeobox gene Nkx2-5. *Gene Dev* **9:** 1654–1666.

Mably JD, Liew CC, 1996. Factors involved in cardiogenesis and the regulation of cardiac-specific gene expression. *Circ Res* **79:** 4–13.

Harel-Bellan A, 1998. Retinoblastoma protein represses transcription by recruiting a histone deacetylase. *Nature* **391:** 601–605.

McPherson CE, Shim E-Y, Friedman DS, Zaret KS, 1993. An active tissue-specific enhancer and bound transcription factors existing in a precisely positioned nucleosomal array. *Cell* **75:** 387–398.

Mizzen CA, Yang X-J, Kokubo T, Brownell JE, Bannister AJ, Owen-Hughes T, Workman J, Wang L, Berger SL, Kouzarides T, Nakatani Y, Allis CD, 1996. The TAFII250 subunit of TFIID has histone acetyltransferase activity. *Cell* **87:** 1261–1270.

Molkentin JD, Markham BE, 1993. Myocyte-specific enhancer-binding factor (MEF-2) regulates α-cardiac myosin heavy chain gene expression *in vitro* and *in vivo*. *J Biol Chem* **268:** 19510–19512.

Molkentin JD, Markham BE, 1994. An M-CAT binding factor and an RSRF-related A-rich binding factor positively regulate expression of the α-cardiac myosin heavy-chain gene *in vivo*. *Mol Cell Biol* **14:** 5065–5065.

Molkentin JD, Brogan RS, Jobe SM, Markham BE, 1993. Expression of the a-myosin heavy chain gene in the heart is regulated in part by an E-box dependent mechanism. *J Biol Chem* **268:** 2602–2609.

Molkentin JD, Kalvakolanu DV, Markham BE, 1994. Transcription factor GATA-4 regulates cardiac muscle- specific expression of the a-myosin heavy-chain gene. *Mol Cell Biol* **14:** 4947–4957.

Molkentin JD, Lin Q, Duncan SA, Olson EN, 1997. Reqirement of the transcription factor GATA4 for heart tube formation and ventral morphogenesis. *Genes Dev* **11:** 1061–1072.

Molkentin JD, Lu J-R, Antos CL, Markham B, Richardson J, Robbins J, Grant SR, Olson E, 1988. A calcineurin-dependent transcriptional pathway for cardiac hypertrophy. *Cell* **93:** 215–228.

Morkin E, 1993. Regulation of myosin heavy chain genes in the heart. *Circulation* **87:** 1451–14661.

MuchardtC, Yaniv M, 1993. A human homologue of Saccharomyces cerevisiae SNF2/SWI2 and Drosophila brm gene potentiate transcriptional activation by the glucocorticoid receptor. *EMBO J* **12:** 4279–4290.

Orgyzko VV, Schiltz RL, Russanova V, Howard BH, Nakatani Y, 1996. The transcriptional coactivators p300 and CBP are histone acetyltransferases. *Cell* **87:** 953–959.

Orphanides G, Lagrange T, Reinberg D, 1996. The general transcription factors of RNA polymerase II. *Genes Dev* **10:** 2657–2683.

Peterson CL, 1996. Multiple SWItches to turn on chromatin. Curr Opin. *Genet Dev* **6:** 171–175.

Peterson CL, Tamkun JW, 1995. The SWI/SNF complex: a chromatin remodelling machine? *Trends Biochem Sci* **20:** 143–146.

Richard-Foy H, Hager GL, 1987. Sequence-specific positioning of nucleosomes over the steroid-inducible MMTV promoter. *EMBO J* **6:** 2321–2328.

Schmid A, Fascher K-D, Horz W, 1992. Nucleosome disruption at the yeast PHO5 promoter upon PHO5 induction occurs in the absence of DNA replication. *Cell* **71:** 853–864.

Simpson RT, 1986. Nucleosome positioning *in vivo* and *in vitro*. *BioEssays* **4:** 172–176.

Struhl K, 1998. Histone acetylation and transcriptional regulatory mechanisms. *Genes Dev* **12:** 599–606.

Subramanian A, Gulick J, Neumann J, Knotts S, Robbins J, 1993. Transgenic analysis of the thyroid-responsive elements in the a-cardiac myosin heavy chain gene promoter. *J Biol Chem* **268:** 4331–4336.

Svaren J, Horz W, 1996. Regulation of gene expression by nucleosomes. *Curr Opin Genet Dev* **6:** 164–170.

Taylor ICA, Workman JL, Schuetz TJ, Kingston RE, 1991. Facilitated binding of GAL4 and heat shock factor to nucleosomal templates: differential function of DNA-binding domains. *Genes Dev* **5:** 1285–1298.

Thompson CM, Young RA, 1995. General requirement for RNA polymerase II holoenzymes *in vivo*. *Proc Natl Acad Sci USA* **92:** 4587–4590.
Thompson WR, Nadal-Girard B, MahdaviV A, 1991. MyoD1-independent muscle-specific enhancer controls the expression of the b-myosin heavy chain gene skeletal and cardiac muscle cells. *J Biol Chem* **266:** 22678–22688.
Travers AA, 1987. DNA bending and nucleosome positioning. *Trends Biochem Sci* **12:** 108–112.
Truss M, Bartsch J, Schelbert A, Hache JG, Beato M, 1995. Hormone induce binding of receptors and transcription factors to a rearranged nucleosome the MMTV ptomoter *in vivo*. *EMBO J* **14:** 1737–1751.
Tsukiyama T, Wu C, 1995. Purification and properties an ATP-dependent nucleosome remodelling factor. *Cell* **83:** 1011–1020.
Wallrath LL, Lu Q, Granok H, Elgin SCR, 1994. Architectural variations of inducible eukaryotic promoters: preset and remodelling chromatin structures. *BioEssays* **16:** 165–170.
Wang W, Xue Y, Zhou S, Kuo A, Cairns BR, Crabtree GR, 1996a. Diversity and specialization of mammalian SWI/SNF complexes. *Genes Dev* **10:** 2117–2130.
Wang W, Cote J, Xue Y, Zuo S, Khavari PA, Biggar SR, Muchardt C, Kalpana GV, Goff SP, Yaniv M, Workman JL, Crabtree GR, 1996b. Purification and biochemical heterogeneity of the mammalian SWI-SNF complex. *EMBO J* **15:** 5370–5382.
Waspe LE, Ordahl CP, Simpson PC, 1990. The cardiac b-myosin heavy chain isogene is induced selectively in a1-adrenergic receptor-stimulated hypertrophy of cultured rat heart myocytes. *J Clin Invest* **85:** 1206–1214.
Wilson CJ, Chao DM, Imbalzano AN, Schnitzler GR, Kingston RE, Young RA, 1996. RNA polymerase II holoenzyme contains SWI/SNF regulators involved in chromatin remodelling. *Cell* **84:** 235–244.
Winston F, Carlson M, 1992. Yeast SWI/SNF: transcriptional activators and the SPT/SIN chromatin connection. *Trends Genet* **8:** 387–391.
Winter BB, Arnold HH, 1987. Tissue-specific DNase I-hypersensitive sites and hypomethylation in the chicken cardiac myosin light chain gene (L2-A) *J Biol Chem* **262:** 13750–13757.
WolffeAP, 1997. Transcriptional control: Sinful repression. *Nature* **387:** 16–17.
Wolffe AP, Pruss D, 1996. Targeting chromatin disruption: transcription regulators that acetylate histones. *Cell* **84:** 817–819.

Yang W-M, Yao Y-L, Sun J-M, Davie JR, Seto E, 1997. Isolation and characterization of cDNAs corresponding to an additional member of the human histone deacetylase gene family. *J Biol Chem* **272:** 28001–28007.

Yang XJ, Ogryzko VV, Nishikawa J, Howard BH, Nakatani Y, 1996. A p300/CBP-associated factor that competes with the adenoviral oncoprotein E1A. *Nature* **382:** 319–324.

Zeller R, Bloch KD, Williams BS, Arceci RJ, Seidman CE, 1987. Localized expression of the atrial natriuretic factor gene during cardiac embryogenesis. *Genes Dev* **1:** 693–698.

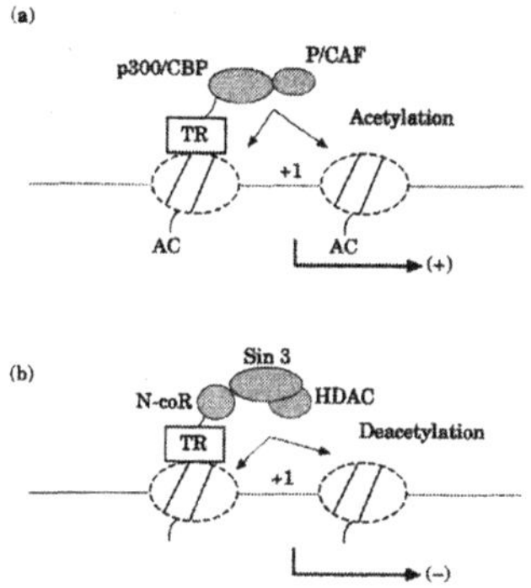

Figure 1 Thyroid hormone receptor recruits histone acetyltransferase or histone deacetylase to modify the nucleosomal structure. (a) Upon ligand activation, thyroid hormone receptor (TR) recruits its co-activator, p300/CBP (with P/CAF), to the vicinity of promoter region. The p300/CBP and P/CAF are able to acetylate the histone tail (AC), thereby relaxation of the histone and DNA interaction. (b) On the repressor element, TR recruits the histone deacetylase via its co-repressor N-CoR and Sin3 complex to turn off gene expression.

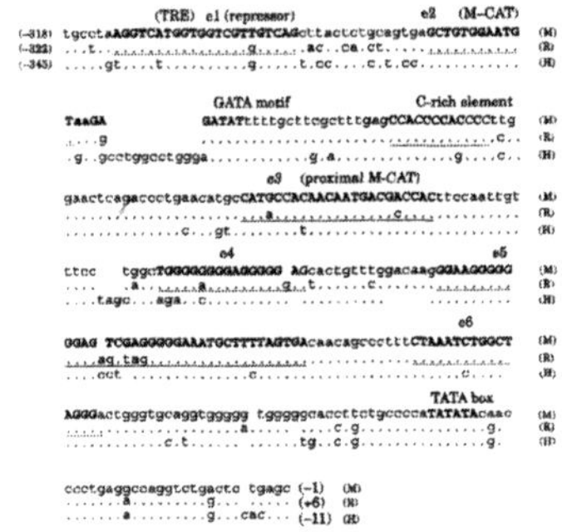

Figure 2 The cardiac β-MyHC gene promoter is highly conserved across species. M, hamster; R, rat; H, human. The be1 to be6 elements are reported by Thompson *et al.* (1991). The M-CAT motif of human and rat β-MyHC gene was identified by Flink *et al.* (1992) and by Kariya *et al.* (1993), respectively.

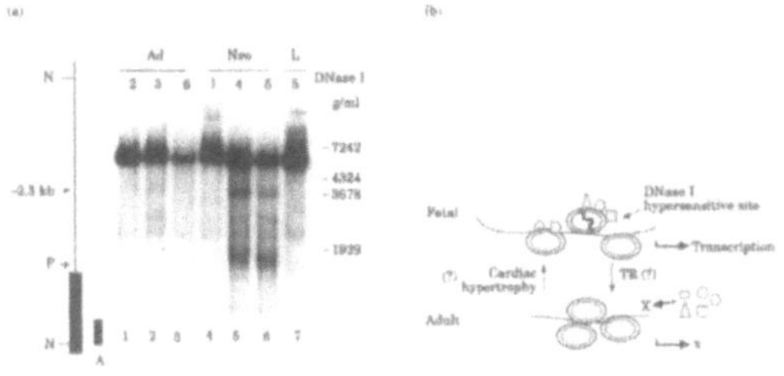

Figure 3 Chromatin remodelling of the cardiac b-MyHC gene. (a) Nuclei from neonatal heart (Neo, lanes 4–6), neonatal liver (L, lane 7), and adult heart (Ad, lanes 1–3) were treated with increasing concentrations of DNase I at 37°C for 2 min, as indicated at the top of the gel. Purified genomic DNA (25lg) were digested with *Nco*I (N), electrophoresed on a 1% agarose gel, and subjected to Southern blot hybridization with probe A (700bp) whose position on the β-MyHC gene is also indicated. The vertical thin line and bold line represent the upstream and coding regions of the β-MyHC gene, respectively. Undigested or partially digested DNAs are observed on the lanes 4 and 7, in the position higher than that of genomic fragment (7018bp), *Nco*I (-5269) to *Nco*I (+1722). P: promoter DHS. This figure is adopted from Huang and Liew (1998). (b) In fetal stage, the proximal promoter region (+1 to-294) of the cardiac β-MyHC gene is organized in three specifically-positioned nucleosomes. The second nucleosome (the transparent circle) is DNase I hypersensitive, allowing all transcription factors to bind to their binding sites. The DNase I hypersensitivity of the β-MyHC gene promoter is lost in adult stage, which is probably caused by the effect of TR. The formation of compacted nucleosome structure excludes the transcription factors from the β-MyHC gene promoter. In addition to *cis*-acting elements and *trans*-acting factors, chromatin structure may also play a crucial role during the reactivation of the fetal genetic program in the hypertrophic heart.

Glossary

actant* Can be thought of as that which accomplishes or undergoes an act. The term is linked with a particular conception of the relations between sentences and the elements of their construction (syntax) which interrelates the functions of the elementary utterance. The functions, such as subject, object, predicate are defined independently of their realization in syntagmatic units (see below).

allele One of the possible mutational states of a gene, distinguished from the other alleles by effects that are seen visibly (phenotypically).

anti-subject* Contingent upon a narrative schema first defined as polemical, implying its presence (adjacent to, or rather opposite to) the subject, which it has to confront.

central dogma The concept that information flow progresses from DNA to RNA to proteins. Although exceptions are known, this idea is central to an understanding of gene function.

chromatin/chromosomes A DNA molecule complexed with RNA and proteins to form a threadlike structure containing genetic information arranged in linear sequence. There are 23 pairs of chromosomes in the human genome. Chromatin is the term used to describe the complex of DNA, RNA histones and nonhistone proteins that make up chromosomes.

cis-acting sequences Sequences on the DNA molecule to which molecules, such as transcription factors, bind in order to exert their effects.

codes Hermenetic, Proairetic, Referential, Symbolic, Semic. Organizes the units of a narrative in simultaneous fashion. Each sentence of a story is categorized by one or more of these codes.

competence* Forms a counterpart to performance. It is a knowing how to do, it is "that something" that makes doing possible. Moreover, this knowing-how-to-do, insofar, as it is a potential act is separable from the doing on which it

* Denotes definition directly taken from A.-J. Greimas' *Analytical Dictionary*.

bears. It can be considered a modal structure (see below) that belongs to the study of human action and sets up the subject as actant (see above).

connotation Used variously to refer to differences in meaning that cannot be reduced to differences in literal, or denotative, meaning.

content plane Along with the expression plane accounts of the existence of meaningful utterances The term content is thus synonymous with Saussure's signified.

deep syntax* (deep structures) opposed in semiotics to surface structure. Based on a series of transformations between two analyses of the same sentence, the more simple and abstract of which is located at the deep level. Both deep and surface syntax depend on the model of competence (see below) and in semiotics the use of the dichotomy is necessarily set in the general theory of the generation of signification. The notion of depth being relative, each domain of the generation of discourse refers back to a deeper domain and so on until the deep structure par excellence, which is the elementary structure of signification is reached. It is this point from which the generative trajectory originates.

denotation The relation between a lexical unit and whatever objects it is used to refer to. E.g. bull denotes a class of animals, brown a property of individuals or objects.

discursive level The taking over of semionarrative structures and transforming them into discursive structures. The discourse is the result of this manipulation of the deep forms, which brings about a surplus of signifying articulations. The procedures of setting into discourse have to bridge the gap between the narrative syntax and narrative semantics (which constitute the surface level of semiotic structures.) Discursive syntax is divided into actorialization, spatialization and temporalization.

DNA A macromolecule usually consisting of antiparallel polynucleotide chains held together by hydrogen bonds, in which the sugar residues are deoxyribose. The primary carrier of genetic information.

DNA response elements An element that causes a gene to respond to respond to a regulatory transcription factor. Examples are the heat shock response elements, glucocorticoid response elements and serum response elements.

DNaseI hypersensitivity sites When chromatin is digested with the enzyme DNaseI, the first effect is the introduction of breaks in the duplex at specific sites, called hypersensitive sites. Since susceptibility to DNaseI reflects the availability of DNA in chromatin, these sites represent chromatin regions in which the DNA is particularly exposed because it is not organized in the usual nucleosomal structure.

experiment and theory versus observation and theory The relationships between theory and experiment differ at different stages of development, nor do all of the natural sciences go through the same cycle. According to Popper, "the view that observations, and even more so observation statements, and statements of experimental results, are always interpretations of facts observed; that they are interpretations in light of theories." Hacking argues that much truly fundamental research precedes any relevant theory whatsoever. Ampère insisted that theoretical science was a study of noumena behind the phenomena. Positivism makes a sharp distinction between theory and observation. Observation terms were taken to raise no problems regarding their meanings, since they referred directly to experience. As to the acceptability of assertions, the program was to show how scientific hypotheses were related to empirical evidence verifying or falsifying them (or confirming or disconfirming them); and if there were any other factors besides empirical evidence influencing the acceptability of scientific hypotheses, those other factors, if at all possible, should be characterized in formal terms as rigorously as the concept of verification or confirmation. However, there have been responses to this view. One is the realistic, or conservative, response, that states there is no significant distinction between observable and unobservable entities; the other is the radical, or idealistic, response in which all observation statements are theory-loaded.

expression plane The Saussurian signifier.

figurative plane* When the content corresponds to something on the expression plane. Discursive semantics has a figurative component. For example, consider a discourse-utterance, a subject disjoined from the object, which is but a goal sought by it. The object stands invested with value that is a form of modality (see below) of being able. From there the discourse can move into gear: the narrative program (see below) consists in joining the subject with the value it seeks. The discourse is said to be figurativized at the moment the syntactic object receives semantic investment which will permit the enunciate to recognize it as a figure. The discourse which expresses the quest for the object, the exercise and, possibly at the end, the recognition of another of the power which it allows it to be manifested, is figurative discourse.

gel electrophoresis An experimental technique in which DNA sequences are run out, or resolved, on a polyacrylamide matrix, called a gel. This gel is run in an electrode buffer in which the DNA sequences, negatively charged, run toward a positive electrode. The purpose of running DNA on a gel is to separate the sequences according to size.

gene expression The phenomenon whereby a gene is transcribed into an RNA molecule (see **transcription** and **mRNA**).

genetic code The nucleotide triplets that code for the 20 amino acids or for chain initiation or termination.

genetic polymorphism The stable coexistence of two are more discontinuous genes in a population.

hermeneutics A method of textual analysis, the term means to interpret. Hermeneutics is an artful form of understanding and a process of exposing hidden meanings. Historically, hermeneutics has been associated with the interpretation of biblical texts. Hermeneutics has been traditionally associated with Heidegger, in a branch of hermeneutics known as Heideggerian hermeneutics. Heidegger emphasized the ontology of Being as opposed to the epistemologic question of knowing. A person, a Being-in-the-World, cannot be separated from the world. Heidegger's philosophy offered an alternate world view from Husserl's belief of a subject-object split. As human-being or *Dasien*, our meaning is codeveloped through being born human and through our life experiences and background. Heidegger acknowledged our background as "throwness." Persons are 'thrown" or born into a particular gender, culture, and history. While throwness prohibits an objective viewpoint, it enables persons to have shared practices and common meanings by virtue of the fact that they share a similar kind of *Being*. Heidegger did not believe it was possible to "bracket" our assumptions of the world. Yet, through authentic reflection we may be able to become aware of many of our assumptions. Heideggarian hermeneutics can be defined as a way to "interpret the shared meanings and practices that we have for our experiences within a context."

histone acetylation (histones): The process whereby an acetyl molecule is attached to an histone protein molecule. Histone acetylation is usually a necessary contingent for gene expression or activation.

hypotactic/hyponymic relation* Is meant in linguistics the hierarchical relation linking two terms situated at two different stages of derivation (example: the relation between main and subordinate clauses). As distinguished from hyponymic relation which defines the position of the semic terms of a single category invested within the semiotic square (see below), hypotaxis could be identified, for example, with implication, which is a relation of unilateral presuppositions between one of the prime terms and the negation of the contrary term.

induction Formulated in the tradition of Carnap, induction is whereby an investigator should make precise observations, conduct experiments with care, and honestly record results; then make the generalizations and draw analogies and gradually work up to hypothesis and theories, all the time developing new concepts to make sense of and organize the facts.

isotopy* Designated as iterativity (see below) along a syntagmatic (see below) chain of units which assure the homogeneity of an utterance-discourse. Due to the syntagm nature of the isotopy, two semic categories are necessary for its establishment. The four terms of the semiotic square (see below) are isotopes.

iterativity* Is the reproduction along the syntagmatic axis (see below) of identical comparable entities which are situated at the same level of analysis. Iterativeness is to be distinguished from recursiveness; the latter is characterized by the repetition of the same entities located at different levels of a common hierarchy.

loss of heterozygosity (LOH) When one allele of two different alleles (a genetic condition called heterozygosity) coding for a genotype is lost upon certain genetic processes.

microarray A glass membrane or DNA chip used to assay gene expression. The objective is to determine which genes are expressed in a particular cell or tissue. Gene sequences are immobilized on the membrane and hybridized with labeled complementary DNA. The differentially expressed genes are then determined.

modalities What modifies the predicate of an utterance. The modal values of wanting, having to, being-able, and knowing are modal values that organize discourses semiotically.

mRNA An RNA molecule transcribed from DNA and translated into the amino acid sequence of a polypeptide.

naked DNA The DNA molecule without its associated histone molecules.

narrative program* Is an elementary syntagm (see below) of the surface narrative syntax (see below) composed of an utterance doing governing an utterance of state. The narrative program is to be interpreted as a change of state effected by any subject (S1) affecting any subject (S2)

negative deixis* One of the fundamental dimensions of the semiotic square; by the relation of the implication it links one of terms of the axis of contraries with the contradictory of the other contrary term. A negative deixis is (S) – non(S1).

nucleosome positioning When a DNA sequence is organized into nucleosomes (see below) in only one particular configuration, so that each site on the DNA always is located at particular position on the nucleosome. The regions of DNA attached to the nucleosomes are unique sequences.

nucleosomes A complex of four histone molecules, each present in duplicate, wrapped by two turns of a DNA molecule. One of the basic units of chromosome structure.

packaged DNA The DNA molecule with its associated histone molecules.

paradigmatic axis* Along with syntagmatic, forming a dichotomy. The functions between the entities situated on this axis are correlations (logical disjunctions of the type either…or) while those on the syntagmatic axis (see below) are relations (logical conjuctions of the type "both…and.")

performative* The domain of realization of the competence in its double task of production and interpretation of utterances. Considered as the narrative program (see above) of the competent subject, one that acts (by itself), can serve as a starting point for a semiotic theory of action.

phenomenology A school of philosophical thought which flourished in Germany, particularly in the work of Husserl and of his pupil Heidegger, and was popularized up to a point in France, in the work of Sartre and Merleau-Ponty. It was attempt at the accurate description of consciousness, and at removing once and for all the division of the subject from the object of human thought. Phenomenology takes subject and object to constitute a whole, and the relation between one and the other to be 'intentional.' The objects of our thought are "intended" by ourselves in different ways, of perceiving, imagining, etc. Consciousness is always consciousness of something, it is a whole divisible only falsely into independent subject and object.

positivism The term "positivism" was invented by the French philosopher Auguste Comte in his *Course of Positive Philosophy* (1830-1842).
Six distinguishing features: (1) An emphasis upon verification or some variant of falsification; (2) Pro-observation: what we can see, feel, and touch provides the best content or foundation for all the rest of our non-mathematical knowledge; (3) Anti-cause: There is no causality in nature, over and above the constancy with which events of one kind are followed by events of another kind; (4) Downplaying explanations: explanations may help organize phenomena, but do not provide any deeper answer to why questions except to say that the phenomena occur in such and such a way; (5) Anti-theoretical entities: Positivists tend to be non-realists, not only because they restrict reality to the observable but also because they are against causes and are dubious about explanations.
Positivism today is usually meant to ascribe to a group of philosophers who formed a famous discussion group in Vienna in the 1920s, known as the Vienna

Circle. Rudolf Carnap and Otto Neurath were among the most famous members, with attendance by Karl Popper and Ludwig Wittgenstein. Logical positivism had an emphasis on logic, meaning and analysis of language. For positivists, opposition to unobservable entities goes hand in hand with an opposition to causes.

positive deixis* One of the fundamental dimensions of the semiotic square; by the relation of implication it links one of the contraries with the contradictory of the other contrary term. The positive deixis is (s1) – non(s2).

pragmatism The American philosophy founded by Charles Sanders Peirce (1839-1914) and made popular by William James (1842-1910) and Dewey. Peirce tried to replace truth by method. Truth is whatever is in the end delivered to a community of inquirers who pursue a certain end in a certain way. Peirce did not think that truth is correspondence to the facts: the truths are stable conclusions reached by an unending community of inquirers. Pragmatists are anti-realists and rationality is extrinsic: it is whatever we agree on.
Dewey distinguished his philosophy from that of earlier philosophical pragmatism by calling it instrumentalism, in that things we make are instruments that intervene when we turn experiences into thoughts and deeds that serve our purposes. Theories and laws have no truth in themselves. They are only instruments, not be understood as literal assertions.

realism and rationalism Scientific realism is a general theory of (scientific) knowledge. In one of its forms it assumes that the world is independent of our knowledge-gathering activities and that science is the best way to explore it. Science not only produces predictions, it is also about the nature of things. What is the world? What kinds of things are in it? What is true of them? What is the truth? Realism about entities says that a good many theoretical entities really do exist. Anti-realism denies that, and says that they are fictions, logical constructions, or parts of an intellectual instrument for reasoning about the world. Questions that philosophers of science ask concerning rationalism are: What do we really know? What should we believe? What is evidence? What are good reasons? Is science rational? Carnap favored verification to distinguish science from non-science. Popper urged falsification. Reason was in terms of a theory of confirmation, or in contrast, in method.

segmentation* An operation which isolates textual units and can be considered from the point of view of the generative trajectory procedures which cut the discourse up into parts which establishes the textual units (sentences, paragraphs, chapters, etc.) and sets them out in succession.

semantic level* Is made up of categories that have no relation with the outside world as it is perceived, considered on the abstract level. Furnishes discourse with contextual semes.

semionarrative level* Inventory of semic categories which stemming from the content plane (see above) of natural languages, correspond to figures of expression in natural semiotics. In opposition to the semantic level (see above), together compose the two levels of the signifying universe.

semiotic square* The visual representation of the logical articulation of any semantic category. The elementary structure of signification that rests on the distinction of opposition which characterizes the paradigmatic axis (see above) of language. The terms of the semiotic square, called isotopes, enter into relations on a semantic axis consisting of contradiction or contrariness.

sequences Distinguished from syntagms, designates a textual unit obtained by the procedure of segmentation (see above). Its discrete quality is guaranteed by the presence of demarcators that serve to determine its boundaries. This is done by distinguishing descriptive, dialogue and narrative sequences and in the second, sequences such as "a walk," "a dance," "a dream." Names of the first type aim at constituting discursive units; those of the second type are of a thematic nature.

sign The union of signifier and signified. Or a unit constituted by a semiotic function that is established between entities on the expression plane (or signifier, see below) and on the content plane (or signified, see below) during the language act. The use of language produces, then semiotic manifestation in the form of strings of signs. Can also be referred to as something there in order to represent something else.

signified Constitutes the concept of a sign, can also be referred to as the content plane.

signifier Constitutes the acoustic image of a sign, can also be considered as a whole and incorporating in its articulations the totality of signifieds.

single nucleotide polymorphism(SNP) A point mutation (an aberration in gene sequence) that is carried by some individuals of a population.

structuralism A term broadly used to describe a strain of philosophical thought mainly developed in 20th century France whose analysis concerns structure, an abstract model of organization, including a set of elements and the law of their composition. Even when the nature of these elements varies considerably, what matters is the inner coherence of the whole. What stands out in a structure is that the relationships between the elements are more important than the intrinsic qualities of each element.

subject Refers to a being to an active principle, capable of not only having qualities but of carrying out acts. Within the framework of the elementary utterance, the subject then appears as an actant (see above) the nature of which depends on the function in which it is inscribed.

syntax* In semiotics, syntax and semantics are the two components of semiotic grammar. From the point of view of linguistics, syntax is traditionally considered as one of the two parts (the other being morphology) that constitute grammar. Morphology is the study of units that make up sentences, whereas syntax seeks to describe their relations and/or to establish the rules of their construction.

surface syntax* (surface structure) Can only be defined with respect to deep structures (see above), and a surface sentence is the form that results from a transformation carried out upon its deep organization.

syntagmatic axis* In opposition to the paradigmatic axis (see above, which is defined by relations of the type "either…or") that identifiable entities have among themselves, the syntagmatic axis is characterized by a network of relations of the type "both…and." Consequently, syntagmatics appear as a hierarchy of relations.

syntagms* Used to refer to a combination of elements co-present in an utterance (sentence or discourse). These elements can be recognized because they have a "both...and" type of relation; they can be further defined by the hypotactic relation (see above) that links them to the surperior unit they constitute. The syntagms are recognized by the segmenting of the syntagmatic string, since the establishment of relations between the parts and the segmentable totalities results in the transformation of this string into a syntagmatic hierarchy (see above). Syntagmatic analysis is completed when the last constitutive elements of a syntagm are no longer segmentable and can no longer be considered as syntagms. Paradigmatic analysis then takes over where syntagmatic analysis leaves off.

trans-acting sequences Transcription factors (see below) that bind to sequences on the DNA molecule.

transcription Transfer of genetic information from DNA by synthesis of an RNA molecule copied from a DNA template.

transcription factors (auxiliary factors) Protein molecules that assist RNA polymerase in transcribing mRNA from DNA.

translation The derivation of the amino acid sequence of a polypeptide from the base sequence of an mRNA molecule in association with a ribosome.

Bibliography

Amrine, Frederick, ed. *Literature and Science as Modes of Expression.* Boston: Kluwer Academic Publishers, 1989.

Anderson, Myrdene and Merrell, Floyd, eds. *On Semiotic Modeling.* New York: Mouton de Gruyter, 1991.

Aronson, Jerrold. *A Realist Philosophy of Science.* New York: St. Martin's Press, 1984.

Bal, Mieke. *Narratology: Introduction to the Theory of Narrative.* Toronto: University of Toronto Press, 1997.

Barnes, Barry. *T.S. Kuhn and Social Science.* London: Macmillan, 1982.

———. "Thomas Kuhn," in *The Return of Grand Theory in the Human Sciences*, ed. Quentin Skinner, 83-100. Cambridge: Cambridge University Press, 1985b.

Barthes, Roland. *Image Music Text.* Trans. Stephen Heath. New York: Hill and Wang, 1977.

———. *Mythologies.* Trans. Annette Lavers. New York: Hill and Wang, 1972.

———. *S/Z.* Trans. Richard Miller. New York: Hill and Wang, 1974.

———. *Writing Degree Zero.* Trans. Annette Lavers and Colin Smith. New York: Hill and Wang, 1967.

Bastide, Françoise. *La Démonstration. Documents.* **3** (1-38) 1981.

Bloor, David. *Wittgenstein: A Social Theory of Knowledge* London: Macmillan, 1983.

Bonner, Stephen E. *Of Critical Theory and Its Theorists.* London: Routledge, 2002.

Britton, Bruce and Pellegrini, Anthony, eds. *Narrative Thought and Narrative Language.* London: Lawrence Erlbaum Associates., 1990.

Boyd, R. "Metaphor and Theory Change: What is 'Metaphor' For?" in *Metaphor and Thought*, ed. A. Ortony, 356-408. Cambridge: Cambridge University Press, 1979.

Brown, T.A. *Genomes.* New York: BIOS Scientific Publishers, Ltd. 1999.

Burley, Jonathan and Harris, John, eds. *A Companion to Genethics.* Malden: Blackwell Publishing, 2002.

Callon, Michel and Latour, Bruno. "Don't Throw out the Baby with the Bath School! A Reply to Collins and Yearly." In Pickering, ed., 343-368, 1992.

Caws, Peter. *The Philosophy of Science*. Princeton: Van Nostrand Company, 1965.

Childs, B. and Valle, D. *Genetics, Biology and Disease. Annual Review of Genomics and Human Genetics* 01 (2000) 1-19.

Cobley, Paul and Jansz Litza. *Introducing Semiotics*. Cambridge: Icon Books Ltd.,1999.

Collins, Harry. *Changing Order: Replication and Induction in Scientific Practice*. Chicago: University of Chicago Press. 1985.

Derrida, Jacques. *Of Grammatology*. Trans. Gayatri Spivak. Baltimore: Johns Hopkins University Press, 1976.

———. *Writing and Difference*. Trans. Alan Bass. Chicago: University of Chicago Press, 1972.

Descartes, René. *Discourse on Method and Meditations on First Philosophy*. Trans. Donald Cress. Indianapolis: Hackett Publishing Company, 1637, 1641.

Doyle, Richard. *On Beyond Living: Rhetorical Transformations of the Life Sciences.* Stanford: Stanford University Press, 1997.

Eco, Umberto. *A Theory of Semiotics*. Bloomington: Indiana University Press, 1979.

———. *Semiotics and the Philosophy of Language*. Bloomington: Indiana University Press, 1986.

Feyerabend, Paul. *Against Method.* New York: Verso, 1975.

———. *Realism, Rationalism and Scientific Method: Philosophical Papers, vol. 1*. Cambridge: Press Syndicate of University of Cambridge, 1981.

Frawley, W. "Relational Models and the Philosophy of Science" in *Relational Models of Lexicon*, ed. M. Evens, 335-72. Cambridge: Cambridge University Press, 1988.

Gert, Bernard, ed. *Morality and the New Genetics*. Boston: Jones and Bartlett Publishers, 1996.

Golinski, Jan. *Making Natural Knowledge: Constructivism and the History of Science.* Cambridge: Cambridge University Press, 1998.

Gooding, David. "Putting Agency Back into Experiment," in *Science as Practice and Culture*, Andrew Pickering, ed. Chicago: University of Chicago Press, 1992.

Greimas, A.J. *Maupassant: The Semiotics of Text.* Trans. Paul Perron. Philadelpha: John Benjamins Publishing Company, 1988.

———. *On Meaning: Selected Writings in Semiotic Theory*. Trans. Paul J. Perron and Frank H. Collins. Minneapolis: University of Minnesota Press, 1987.

———. *Structural Semantics: An Attempt at a Method.* Lincoln: University of Nebraska Press, 1983.

Greimas, A.J. and Courtés, J. *Semiotics and Language: An Analytical Dictionary*. Bloomington: Indiana University Press, 1979.

Hacking, Ian. *Representing and Intervening: Introductory Topics in the Philosophy of Natural Science*. Cambridge: Cambridge University Press, 1983.

Hacking Ian, ed. *Scientific Revolutions*. Oxford: Oxford University Press, 1981.

Harari, Josué, V., ed. *Textual Strategies: Perspectives in Poststructuralist Criticism*. Ithaca: Cornell University Press, 1979.

Harré, Rom. "Some Narrative Convention of Scientific Discourse," in *Narrative in Culture: The Use of Storytelling in the Sciences, Philosophy and Literature*, ed. Cristopher Nash. New York: Routledge, 1991.

Hawkes, Terence. *Structuralism and Semiotics*. London: Routledge. 1991.

Heidegger, Martin. *Being and Time: A Translation of Sein and Zeit*, Trans. by Joan Stambaugh. Albany: State University of New York Press, 1997.

Hempel, Carl. *Philosophy of Natural Science*. Upper Saddle River: Prentice Hall, 1966.

Herman, David, ed. *Narrative Theory and the Cognitive Sciences*. Stanford: CSLI Publications, 2003.

———. *Narratologies*. Columbus: Ohio State University Press, 1999.

Hoffmeyer, Jesper and Emmeche, Claus. "Code-duality and the semiotics of nature". pp. 117-166 in: *On Semiotic Modelling* , Myrdene Anderson and Floyd Merrell, (eds.). Mouton de Gryter, Berlin, 1991.

Huang, Weei-Yuarn and Liew Choong-Chin. *Chromatin Structure and Gene Expression*. Journal of Molecular and Cellular Cardiology (1998) 30: 1673-1681.

Husserl, Edmund. *On the Phenomenology of the Consciousness of Internal Time (1893-1917)*. Translated by John Barnett Brough and edited by Martin Heidegger. Collected Works: Volume 4. The Hague, Netherlands: Kluwer Academic Publishers, 1991.

Jahn, Manfred. " 'Awake! Open your eyes!' The Cognitive Logic of External and Internal Stories in *Narrative Theory and the Cognitive Sciences*, ed. David Herman, 195-213. Stanford: CSLI Publications, 2003.

Jakobson, Roman. *On Language*. Cambridge: Harvard University Press, 1995.

Johnson, Barbara. *The Critical Difference: Essays in the Contemporary Rhetoric of Reading*. Baltimore: Johns Hopkins University Press, 1980.

Kay, Lily. "A Book of Life? How a Genetic Code Became a Language," in *Controlling Our Destinies: Historical, Philosophical, Ethical and Theological Perspectives on the Human Genome Project,* ed. Philip Sloan. Notre Dame: Notre Dame Press, 2000.

Kent, Theodore. *Mapping the Human Genome: Reality, Morality and Deity.* Lanham: University Press of America, 1995.

Kevles, Daniel and Hood Leroy, eds. *The Code of Codes: Scientific and Social Issues in the Human Genome Project.* Cambridge: Harvard University Press, 1993.

Kitcher, Philip. *The Advancement of Science.* Oxford: Oxford University Press, 1993.

Klemke, E.D., et al., eds. *Introductory Readings in the Philosophy of Science.* Amherst: Prometheus Books, 1998.

Knorr-Cetina, Karin. *Epistemic Cultures: How the Sciences Make Knowledge.* Cambridge: Harvard University Press, 1999.

Knudson, Suzanne. "Communicating novel and conventional scientific metaphors: a study of the development of the metaphor of the genetic code" in Public Understanding of Science, 373-392, Vol. 14, 2005.

Kristeva, Julia, ed. *Essays in Semiotics.* Paris: Mouton, 1971.

Kuhn, Thomas. *The Structure of Scientific Revolutions,* 3rd edition. Chicago: University of Chicago Press, 1996.

Lakatos, Irme. "History of Science and Its Rational Reconstructions" in *Scientific Revolutions,* ed. Ian Hacking. Oxford: Oxford University Press, 107-127, 1981.

Latour, Bruno and Woolgar, Steve. *Laboratory Life: The Construction of Scientific Facts.* Princeton: Princeton University Press, 1986.

Latour, Bruno. *Science in Action.* Cambridge: Harvard University Press, 1987.

Laudan, Larry. *Progress and Its Problems: Towards a Theory of Scientific Growth.* Berkeley: University of California Press, 1977.

Levine, George, ed. *One Culture: Essays in Science and Literature.* Madison: University of Wisconsin Press, 1987.

Lewin, Benjamin. *Genes VII.* Oxford: Oxford University Press, 2000.

Lomvarda S. and Thanos, D. *Nucleosomal sliding via TBP DNA binding in vivo. Cell,* (2001) Sep 21 106(6): 685-696.

Lyotard, Jean-François. *The Postmodern Condition.* Minneapolis: University of Minnesota Press, 1979.

Margolin, Uli. *Shifted Temporal Perspective in Narrative*: An article from: Narrative [HTML]. Cincinnati: Ohio State University Press, 2001.

Martin, Wallace. *Recent Theories of Narrative*. Ithaca: Cornell University Press, 1986.

Matthews, P.H. *Oxford Concise Dictionary of Linguistics*. Oxford: Oxford University Press, 1997.

Maxwell, Donald. *Science or Literature? The Divergent Cultures of Discovery and Creation*. New York: Peter Lang Publishing, 2000.

McHoul, Alec. *Semiotic Investigations: Towards an Effective Semiotics*. Lincoln: University of Nebraska Press, 1984.

Morgan, Griffith. "Metaphor and Model: Common Ground Between Literature and Science" in *Semiotics As A Bridge Between the Humanities and the Sciences*, eds. Perron, Paul, Sbrocchi, Leonard, Colilli, Paul, Danesi, Marcel, 472-478. Brooklyn: LEGAS, 2000.

Myers, Greg. "Making a Discovery: Narratives of Split Genes, in_*Narrative in Culture: The Use of Storytelling in the Sciences, Philosophy and Literature*, ed.
Cristopher Nash. New York: Routledge, 1991.

Nuessel, Frank. "The Use of Metaphor to Comprehend and Explicate Scientific Theory" in *Semiotics As A Bridge Between the Humanities and the Sciences* eds. Perron, Paul, Sbrocchi, Leonard, Colilli, Paul, Danesi, Marcel, 479-500. Brooklyn: LEGAS, 2000.

Norris, Christopher. *Derrida*. Cambridge: Harvard University Press, 1987.

Olby, Robert. *The Path to the Double Helix: The Discovery of DNA*. New York: Dover Publications, Inc., 1994.

Oreskes, Naomi, ed. *Plate Tectonics: An Insider's History of the Modern Theory of the Earth*. Boulder: Westview Press, 2003.

———. *The Rejection of Continental Drift*. Oxford: Oxford University Press, 1999.

Ormiston, Gayle L. and Sassower, Raphael. *Narrative Experiments: The Discursive Authority of Science and Technology*. Minneapolis: University of Minnesota Press, 1989.

Paduceva, Elena. "La langue de la logique mathematique comme modèle sémantique pour la langue naturelle." In Kristeva, ed. 175-190, 1971.

Perron, Paul, Sbrocchi, Leonard, Colilli, Paul, Danesi, Marcel, eds. *Semiotics As A Bridge Between the Humanities and the Sciences*. Brooklyn: LEGAS, 2000.

Peterfreund, Stuart, ed. *Literature and Science: Theory and Practice*. Boston: Northeastern University Press, 1990.

Peirce, Charles Sanders. *Writing on Semiotic*. Chapel Hill: University of North Carolina Press, 1991.

Pickering, Andrew, ed., *Science as Practice and Culture.* Chicago: University of Chicago Press, 1992.

Popper, Karl. *The Logic of Scientific Discovery.* Cornwall: Routledge, 1959.

———. "The Rationality of Scientific Revolutions." In Hacking, ed., 80-106, 1981.

Porter-Abbot, H. *The Cambridge Introduction to Narrative.* Cambridge: Cambridge University Press, 2002.

Poster, Mark. *Critical Theory and Poststructuralism.* Ithaca: Cornell University Press, 1989.

Putnam, Hilary. "The Corroboration of Theories." In Hacking, ed., 60-79, 1981.

Pugh, B. Franklin. *Control of gene expression through regulation of the TATA-binding protein.* Gene (2000) **255:** 1-14.

Rouse, Joseph. *The Narrative Reconstruction of Science* Inquiry 33:179-196, 1990.

Ruse, Michael, ed. *Philosophy of Biology.* Amherst: Prometheus Books, 1998.

Saussure, Ferdinand de. *Course In General Linguistics.* Trans. Roy Harris. Peru: Open Court Publishing, 1986.

Schaff, Adam. "L'objectivité de la connaissance à l lumière de la sociologie de la connaissance et de l'analyse du language." In Kristeva, ed. 63-94, 1971.

Schleifer, Ronald. *A.J. Greimas and the Nature of Meaning: Linguistics, Semiotics and Discourse Theory.* Kent: Croom Helm, Ltd., 1987.

Schleifer Ronald, Con Davis, Robert, Mergler, Nancy. *Culture and Cognition: The Boundaries of Literary and Scientific Inquiry.* Ithaca: Cornell University Press, 1992.

Sebeok, Thomas. "Semiotics as Bridge Between Humanities and Sciences" in *Semiotics As A Bridge Between the Humanities and the Sciences* eds. Perron, Paul, Sbrocchi, Leonard, Colilli, Paul, Danesi, Marcel, 76-100. Brooklyn: LEGAS, 2000

Sewack, Gerald F., Ellis, Thomas W., Hansen Ulla. *Binding of TATA Binding Protein to a Naturally Positioned Nucleosome is Facilitated by Histone Acetylation.* Molecular and Cellular Biology (2001) 21:1404-1415.

Shapere, Dudley. "Meaning and Scientific Change." In Hacking, ed., 28-59, 1981.

Shapin, Steven and Schaffer, Simon. *Leviathon and the Air Pump: Hobbes, Boyle and the Experimental Life.* Princeton: Princeton University Press, 1986.

Sturrock, John. *Structuralism.* Malden: Blackwell Publishing, 2003.

Wilson, David and Bowen, Zack. *Science and Literature: Bridging the Two Cultures.* Gainesville: University Press of Florida, 2001.

Yokomori, Kyoko, Verruzer, Peter C., Tjian Robert. *An interplay between TATA-box binding protein and transcription factors IIE and IIA modulates DNA binding and transcription.* Proc. Natl. Acad. Sci., U.S.A., (1998) 95: 6722-6727.

http://cnx.org/content/m11320/latest/histone.gif

http://www.accessexcellence.org/RC/VL/GG/images/nucleosome.gif

Index

–T–

–W–

BERKELEY INSIGHTS IN LINGUISTICS AND SEMIOTICS

Irmengard Rauch
General Editor

Through the publication of groundbreaking scholarly research, this series deals with language and the multiple and varied paradigms through which it is studied. Language as viewed by linguists represents micrometa-approaches that intersect with macrometa-approaches of semiotists who understand language as an inlay to all experience. This data-based series bridges study of the sciences with that of the humanities.

To order other books in this series, please contact our Customer Service Department at: